A Note to the Teacher

This booklet contains worksheets entitled Life Skills, Evaluating Media Messages, and Ethical Issues in Health. These activities are designed to enable students to apply the knowledge they have gained from the textbook to real-life situations. These worksheets and activities further develop topics that are covered by features in the Pupil's Edition. In the back of this booklet you will find an answer key that gives answers and support to Life Skills Worksheets.

Life Skills

Life Skills Worksheets are designed to give students more practice in certain skills. Each worksheet is labeled according to the skill that it is designed to utilize. These skills are:

- Solving Problems
- Coping
- Setting Goals
- Resisting Pressure
- Assessing Your Health
- Using Community Resources
- Communicating Effectively
- Being a Wise Consumer
- Making Responsible Decisions
- Practicing Self-Care
- Intervention Strategies

Evaluating Media Messages

The messages conveyed in advertisements and other media have a profound influence on our lives. They especially have an impact on the lives of teenagers. We are surrounded by messages about what is desirable and what is undesirable in our culture. Sometimes these messages portray unrealistic images of life or uphold unrealistic standards. In the Evaluating Media Messages activities, students are asked to analyze messages in the media and determine how these messages can influence their health.

Ethical Issues in Health

Decisions regarding today's health issues are not always clear-cut or easy to make. Our values and beliefs play important roles in shaping the decisions we make. Ethical Issues in Health takes the decision-making process to a higher level as students move from examining personal decisions to looking at societal decisions related to health. Students use the activities to analyze ethical issues in a systematic way and draw conclusions based on the thorough analysis of an issue.

Answer Key

Located in the back of this booklet, the Answer Key provides answers to the questions and assignments in the Life Skills Worksheets. Many of the questions and assignments in the Life Skills Worksheets ask for students' opinions or personal experiences and do not have correct answers. In these cases, the Answer Key may provide sample answers or suggest ways to evaluate student effort. The Answer Key also provides background material for those worksheets that handle sensitive subjects or require additional support. The Answer Key does not provide answers for Evaluating Media Messages or Ethical Issues in Health because these activities are based entirely on student opinion and personal experiences.

Table of Contents

20/20

Activity A: Assessing Your Health

Determining Your Health History

Family history may affect your risks for certain diseases. For this reason, it is useful to know about your parents' and grandparents' health histories and to keep a record of your own. Fill out the form below with the help of a parent or guardian.

1.

	Names	Present age or age at death	Major illnesses	Cause of death
Grandmothers				
Grandfathers				
Mother				
Father				
Brothers/Sisters				

2. Your illnesses	Date (or age)	3. Your surgeries	Date (or age)

4. Your immunizations	Date (or age)	5. Your allergies	

Chapter 2

Activity A: Resisting Pressure

Making Immediate Decisions

Sometimes it is important to make a good decision rapidly. Read the passage below and answer the following questions as quickly as possible.

Imagine that you are at a party at your friend Rhonda's house. You are having a good time until you look at your watch and realize that you should have called your parents five minutes ago. They said that you could come to this party if you promised to call them at 10:00 P.M. You go to use the phone, but Rhonda is using it and seems to be ignoring you. Your friend Jack offers to give you a ride on his motorcycle to the nearest phone booth, which is a few blocks away. Jack appears to be a safe driver, but he does not have a helmet for you to wear. You get the feeling that declining Jack's offer will make you seem silly to your friends. After all, its only a couple of blocks.

1. What options do you have? Briefly list the benefits and consequences of each option.

____

____

____

2. What are your values regarding your safety, your parents, and your peers?

____

____

3. What have you decided to do? Finish the story showing the best possible results of this decision.

____

____

4. Now finish the story showing the worst possible results of your decision.

____

____

____

5. Overall, are you happy with your decision? Why or why not?

____

Activity A: Setting Goals

Improving Your Fitness

1. Choose one area of your physical fitness that you would like to improve.

____

2. What exercise would be a good indicator of your fitness in this area? (For example, the number of miles you could run would be an indicator of your endurance.)

____

3. Do you wish to gain endurance or strength in this exercise?

____

4. How could you measure your progress?

____

5. What level of fitness are you currently able to achieve in this exercise?

____

6. What level would you like to achieve?

____

7. Outline below a six-week plan to achieve your goal. Be specific about the kind of exercise and the length of time you will spend exercising. Include the number of repetitions and your goals for increasing repetitions each week. Use additional paper if necessary.

____

____

____

8. Check with your teacher to evaluate whether your goals are realistic. If they are not, explain how you will modify them.

____

____

Chapter

3

Activity B: Practicing Self-Care

Keeping a Sleep Journal

1. Over the next week, fill in the following table. The first four items relate to the conditions of your sleep, and the last four refer to your experience the following day. Keep a journal describing any conditions that affected the quality of your sleep.

	Day 1	Day 2	Day 3	Day 4	Day 5	Day 6	Day 7
Did you eat within 3 hours of bedtime?							
Level of stress at bedtime (1 to 5, 5 = most stress)							
Hours slept							
Number of awakenings							
Restedness upon rising (1 to 5, 5 = most rested)							
Energy level (1 to 5, 5 = most energetic)							
Mood (1 to 5, 5 = best mood)							
Alertness (1 to 5, 5 = most alert)							

2. Do you notice any patterns in your table or journal? If so, why might they exist?

3. Based on your study, what changes could you make to get better sleep?

Chapter

4

Activity A: Being a Wise Consumer

Examining Food Labels

On food labels, ingredients are listed in descending order of weight, but that can be deceptive. For example, fat, sodium, or sugar may be added in several different forms.

Suppose, for example, a jar of strawberry preserves contained four ounces of strawberries, three ounces of cane sugar, and two ounces of corn syrup. The label would read

Ingredients: Strawberries, cane sugar, corn syrup

The consumer would probably assume that the preserves contained more fruit than sugar, but it really contained five ounces of sugar (cane sugar and corn syrup) and four ounces of strawberries.

Fats can also be listed separately. For example, fat may be added as liquid soybean oil, partially hydrogenated corn oil, partially hydrogenated cotton seed oil, and so forth.

Sodium can be added in many forms that are also listed separately. For example, sodium is present in sodium chloride (table salt), monosodium glutamate, sodium bicarbonate (baking soda), and sodium nitrate.

Visit a supermarket and gather a number of food containers or labels. Find five products that list more than one type of sugar, fat, or sodium. List the products and their ingredients below.

____

____

____

____

____

____

____

____

____

____

____

Activity B: Practicing Self-Care

Planning a Menu

You have read about nutrients, food groups, and the recommendations of the Food and Drug Administration. Now apply them to your daily life by creating a menu for the day. Plan for three meals and at least one snack.

Review the Dietary Guidelines for Americans (Figure 4-13), Daily Reference Values (Figure 4-14), and Recommended Daily Intakes (Figure 4-15). Then refer to a table of nutrient composition of foods from the library. Plan a menu that will meet your nutritional requirements while falling within these guidelines. Keep the food pyramid in mind as you plan your menu.

BREAKFAST

LUNCH

SNACK

DINNER

Activity C: Being a Wise Consumer

More for Your Money

Gather the food labels and prices from several types of ready-to-eat breakfast cereals. Also gather labels from several packages of whole grains for cooking, such as oatmeal, cornmeal, cracked wheat, or brown rice.

1. Compare the price of a standard serving of one regular cooked hot cereal, one instant hot cereal, and one dry cold cereal with that of one raw serving of a whole-grain product. List the brand names, main ingredients, and prices below.

____

____

____

____

____

____

2. Using the information from the labels, compare the four products that you chose above. Do not include added milk in your calculations. Which contains more of the following nutrients?

   a. protein   c. vitamin A   e. carbohydrates
   b. fiber   d. fat   f. vitamin B_2

____

____

____

____

____

Chapter 4

Activity D: Assessing Your Health

Evaluating Your Diet

1. List all the foods you have eaten in the last two days.

____

____

____

____

____

____

2. Review the dietary guidelines in this chapter and the information about the fat and sugar content of foods. Decide whether or not your diet contains more fat or sugar than the Food and Drug Administration recommends. If it does, list several ways you could reduce your fat or sugar intake.

____

____

____

3. Refer to the RDIs in Figure 4-15 to determine your nutritional requirements. Then refer to a table of nutrient composition of foods from the library. If you have eaten fast food or prepared brand-name foods in the last two days, try to obtain a resource book that supplies nutritional information for these items. Did you obtain the RDI of the specified nutrients? If not, what foods could you substitute to increase your intake of nutrients without increasing your fat and sugar intake?

____

____

____

4. Review any suggestions you made in items 2 and 3. Decide which, if any, you are willing to actually apply in your life. If you are not willing to apply any of them, refer again to this chapter and the table of nutrient composition of foods. Try to find some acceptable ways to improve your diet.

Activity A: Being a Wise Consumer

Comparing Weight-Loss Programs

Suppose you wish to lose weight sensibly, but you feel that you need help doing it. If so, you might consider using one of the weight-control services listed in the telephone book. To make a wise choice, telephone several of the services and ask them the following questions:

	Service A:	Service B:	Service C:
1. What dieting guidelines does your center follow?			
2. What part does exercise play in your weight-loss plan?			
3. What type of counseling does your service offer?			
4. How much does your service cost?			

Activity A: Solving Problems

Practicing Personal Care

1. You and your friends are planning to spend the day at the beach. You arrive at the beach only to discover that you have forgotten your sunscreen. One of your friends offers to lend you some sunscreen with an SPF factor of 8. What could you do to protect your skin and still enjoy the beach? Give three possible solutions.

____

____

____

____

____

____

____

2. Your friend wants you to hike with her in an area that is thick with poison ivy. She is not allergic to poison ivy, but you are. What are two ways you could handle the situation and avoid an allergic reaction?

____

____

____

____

____

____

3. Your friend has come to school without a comb and his hair looks terrible. He wants to borrow your comb or brush. What is wrong with lending your brush or comb? How could you handle this with your friend?

____

____

____

____

____

____

4. Consuela's solution to friends who want to borrow her comb is to keep a lending comb. She does not use it herself. What are the advantages and disadvantages of her solution?

____

____

____

____

____

____

____

Activity B: Making Responsible Decisions

Toothbrush Choices

1. Use the library to find evaluations of toothbrushes in publications such as consumer magazines. What types and brands of toothbrushes are recommended?

2. Cooperate with five other students who have different dentists. Call your dentist's office, and ask the staff to recommend a type of toothbrush. If you do not have a regular dentist, call the nearest school of dentistry. Compare answers with your classmates.

   - What shape do dentists recommend for a toothbrush?

   - What type of bristles are recommended?

   - What other recommendations do dentists make?

   - Do dentists agree on these points?

3. Do consumer magazines' ratings agree with dentists' recommendations?

4. Some toothpastes contain special additives. Examine advertisements for toothpastes. Visit a store, and read the labels of the various brands. Which substances are represented as cavity fighters or plaque inhibitors? List them.

Which brands have their claims supported by the American Dental Association? What are their cavity- or plaque-fighting ingredients?

5. Study consumer magazines' articles about these ingredients and ratings of the brands that contain them. Record your findings.

6. Work with five students who have different dentists. Call your dentist's office. If you do not have a regular dentist, call the nearest school of dentistry. Ask if your dentist suggests a particular brand of toothpaste. Ask for general comments about fluoride products and plaque inhibitors. Record your findings.

7. Choose a toothpaste for yourself.

Activity A: Solving Problems

Making Friends

Perhaps you have many friends where you live now. However, someday you may need to move to an area where you don't know anybody. How could you make friends and keep from being lonely?

1. Assume somebody about 25 years old has moved to your town or city. The person is not in school and is working at a job that does not allow opportunities for meeting people. What would you suggest that this person do to meet people?

____

____

____

____

____

2. Make a list of organizations in your community that new people could join to bring them in contact with others.

____

____

____

____

____

3. List some volunteer activities available in your community. Which would be most likely to allow a young adult to make friends?

____

____

____

____

Activity B: Coping

Defense Mechanisms

Read the passage below and answer the questions that follow.

Defense mechanisms can work against you if you use them instead of recognizing a problem and taking steps to handle it. However, they can also be useful in protecting yourself from unnecessary emotional pain. For example, a high school athlete who received a crippling injury might successfully substitute the goal of becoming a writer for the goal of playing football. On the other hand, a player who always projects his or her athletic failures on team members, blaming them for his or her errors, will be less likely to improve his or her technique.

1. Choose three defense mechanisms, and write examples of somebody using them in a positive, constructive way or as necessary coping mechanisms.

____

____

____

____

____

____

____

2. Choose three different defense mechanisms, and write examples of somebody using them in a counterproductive way.

____

____

____

____

____

____

____

Chapter 7

Activity C: Assessing Your Health

Using Maslow's Hierarchy of Needs

Sometimes it's hard to identify the cause of a bad mood. A person may feel down, frustrated, or irritable without being able to identify what makes him or her feel that way. This happens most often when several problems overlap, making it difficult to see any one problem clearly. Maslow's Hierarchy of Needs can be a useful tool for identifying, untangling, and solving such problems. For example, if you frequently feel irritable and negative in the late afternoon, address your physical needs first. If you are low on energy, having a healthy snack may increase your energy and improve your mood. You will now have more energy and clarity to move on to higher-level needs.

The next time you feel anxious, down, frustrated, or irritable, use this checklist to assess and address your problems. When you begin, describe how you feel on the lines below.

____

____

Part I: Physiological Needs

1. How long ago did you last eat? ______________ Was the meal nutritious? ___________

   If you ate more than three to four hours ago or did not eat well, try having a healthy snack.

2. Did you get enough sleep last night? ___________

   If you answered no, is it possible for you to take a short nap or avoid exerting a great deal of energy? ___________

3. Are you experiencing any physical pain, such as a headache or muscle ache? ___________

   If you answered yes, what are some ways you can lessen this pain?

   ____

   ____

4. Do you still feel bad? ___________

   If yes, you may be experiencing a hormonal imbalance normal in adolescents that could be helped by exercise. Try doing some kind of exercise for 20 minutes.

Part II: Safety and Security

5. Do you still feel bad? ___________

   If so, address your needs for safety and security. Do you feel safe where you are right now? ___________ Are you feeling threatened in any way? ___________

6. If you do not feel completely safe and secure, how could you change your situation so that you do?

   ____

   ____

Life Skills 7C cont.

Part III: Love and Affection

7. Do you still feel bad? _______________

8. Explore your needs for love and affection. Is there someone you would like to resolve a conflict with or need to spend time with? _______________

9. Is there anything you can do right now to get what you need? _______________
If not, write the need down and make a plan to address it at some later time.

Part IV: Self-Esteem

10. Do you still feel bad? _______________

11. Do you feel basically good about yourself and others? _______________
If not, is there something you would like to work on? Explain your answer.

---

12. Make a plan for action, and set a target date for completing your goal. Chapter 8 provides information about how to build self-esteem.

Part V: Self-Actualization

13. Now that you have addressed at least some of your other needs, think about your talents and dreams. Is there an activity that would make you feel more fulfilled? Write down five things you can do to explore and develop your talents. As you do this, realize that self-actualization is a process that continues throughout one's lifetime. Exploring your talents and setting goals is one part of this process.

---

14. Sometimes there are fears, feelings, or beliefs that can prevent us from embracing our goals. Can you identify any obstacles that you feel prevent you from pursuing your goals? List these obstacles and write what you can do to overcome them.

---

15. Describe how you feel now. How does this compare with how you felt before completing this exercise?

Activity D: Using Community Resources

Help in Your Community

At some point, you may wish to seek counseling for yourself or someone you know. Use these activities to become familiar with the options for treatment in your community.

1. Find out what counseling is available through your school or district.

2. List the agencies, other than individual counselors, that provide counseling or treatment for mental and emotional disorders. Look in the telephone book under Hospitals, Counseling, Support Groups, Mental Health, Psychologists, Treatment Centers, Crisis Intervention, and other headings. Call local colleges and ask what they provide. Call available hotlines.

3. Which of these agencies focus on a specific disorder?

4. Which are residential programs?

5. Are any services available without charge or at a reduced rate?

6. Who qualifies for service?

Activity A: Coping

Wise Self-Disclosure

Sharing is important, but certain kinds of self-disclosure can have negative consequences. Decide whether it would be wise for you to confide in each of the following individuals. Write "yes" or "no" on the blank in front of each number.

______ **1.** Jan is upset. She will not tell you what is wrong because she would need to break a confidence to do it.

______ **2.** Ben says he needs to confide in you. Trevor has just confided that he is considering suicide, and Ben doesn't know what to do.

______ **3.** Anita says, "You can talk to me. Reiko thinks you are weird, and she doesn't like you, but I do."

______ **4.** Gary says, "I wonder if anyone else is afraid of asking girls out."

______ **5.** Caesar says, "Do you know how it feels to be left out?"

______ **6.** Donna tells you how much she admires Jack. "Nobody else knows it," she says, "but Jack was addicted to crack at his former school and pulled his life together."

Write how you would respond to your friend in each situation.

____

____

____

____

____

____

____

____

____

Activity B: Using Community Resources

Finding Support Groups

1. Find out what support groups exist in your community for the following people:

   a. Alcoholics

   ____

   b. People in withdrawal from drugs

   ____

   c. People who want to lose or gain weight healthfully

   ____

   d. People with eating disorders

   ____

   e. People who have had a death in the family

   ____

   f. People who are divorced

   ____

2. Find one support group for a different category of people not listed above.

   ____

3. Think of a situation in someone's life that might require a support group. Find out if such a support group exists in your community.

   ____

Activity C: Practicing Self-Care

Choosing Supportive Relationships

As important as it is to make your own decisions about your worth, your self-esteem is sometimes influenced by others around you. For each of the following situations, what advice would you give about preserving self-esteem?

1. Samantha had sex with her boyfriend because he said he would dump her if she would not have sex. Before she and her boyfriend had sex, Samantha hadn't thought seriously about her moral beliefs regarding premarital sex. Now she regrets having had sex. She's worried that her boyfriend will expect to have sex again.

2. Raol and Angel have both asked Sue to a dance. Angel is handsome and popular, but he is always making fun of others. He often teases and embarrasses her. Raol is neither popular nor unpopular. He is sensitive to others' feelings, though, and treats Sue with respect.

3. Francine needs someone to evaluate her term paper before she turns it in. She loves showing her work to Jerry because he always tells her it is excellent. Showing it to Arsenio is sometimes difficult because he does point out flaws. He always suggests ways to improve the paper.

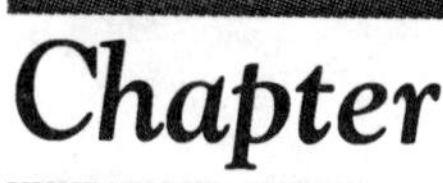

Chapter
8

Activity D: Practicing Self-Care

Evaluating Behaviors

Classify each behavior below by giving it the letter of the appropriate category. Some behaviors may fit in more than one category.

    a. Behaviors tending to lower self-esteem
    b. Crutches that may give a false sense of self-esteem
    c. Sound strategies for building true self-esteem

_______ **1.** Esperanza always lives according to her principles.

_______ **2.** Paul works eight hours a night to afford the fastest car at school.

_______ **3.** Ricardo dates a girl who only wants him for sex.

_______ **4.** Rochelle does what other people tell her to do so they will like her.

_______ **5.** Greg has a best friend who constantly criticizes him.

_______ **6.** Deanna hangs around only with popular people so she will be popular too.

_______ **7.** Martha spends all her money on clothes and spends a lot of time in class fixing her hair and makeup.

_______ **8.** Jasper is taking an advanced ballet class.

_______ **9.** Tran chooses friends he considers inferior to him so he can feel superior by comparison.

_______ **10.** Susan constantly compares herself to other people.

_______ **11.** Jim chooses friends he respects and who respect him.

_______ **12.** Lucille constantly points out others' faults so people will notice how superior she is.

_______ **13.** Leah takes pride in her skill at playing tennis and is always working at improving her game.

Chapter
9

Activity A: Practicing Self-Care

Creating a Relaxation Corner

Where can you go when you feel stressed-out? Sometimes it's best to have an answer to that question before the stress builds up. You can create a relaxation corner in your room or in any other quiet place in your home. First, select a quiet spot away from the main traffic areas of your home. Then, use some of the suggestions listed below for creating an environment that is calm and relaxing for you to spend time in. Try to incorporate things that involve all of the senses. Also, add your own personal touches. After you have finished creating the relaxation corner, draw a picture of it in the space below. Label the things you did and the objects you included to make the area a peaceful place for you.

- a chair with soft cushions, armrests, and a headrest
- a bed or cot with plenty of pillows
- a window for sunlight or a lamp with a soft-light light bulb

- a large picture or poster of a place, such as a beach or forest, that you think is beautiful
- your favorite shells, rocks, or other keepsakes from a peaceful place you like to visit

- a tape- or CD-player and your favorite music
- a musical instrument if you find it relaxing to play music
- a notebook and pencil for writing down your feelings

- a book you find especially funny
- a box full of sweet spices, such as cinnamon or cloves
- incense and an incense burner

Chapter 9

Activity B: Communicating Effectively

How Do You Communicate?

Learning how to communicate effectively with people is a good way to help reduce stress in your life. In the next week, pay special attention to how you communicate with your friends, family, teammates, co-workers, and other important people in your life. Use the following questions to help you evaluate how you communicate:

1. Describe a situation in which you had difficulty trying to communicate with or understand someone this week.

____

____

2. Check all the factors that made communication difficult or ineffective.

_______ I did not attempt to communicate about something that bothered me.

_______ The other person did not attempt to communicate about an issue of concern.

_______ I chose a bad time to try to communicate about the issue.

_______ The person I was communicating with did not understand why I was concerned or upset.

_______ I did not communicate clearly.

_______ The other person did not communicate clearly.

3. Use the list below to help you figure out what you could do to communicate more effectively.

_______ Choose a better time or place to communicate.

_______ State my concerns and feelings clearly and respectfully.

_______ Listen respectfully to the other person.

_______ Let the other person know that I understood his or her concerns.

_______ Ask the other person to give me feedback about his or her perception of what is bothering me.

_______ Work out and implement a solution that is acceptable to both of us.

Chapter 9

Activity C: Coping

Getting Perspective

If low self-esteem is causing you to feel stress, sometimes the best way to relieve the stress and build up your self-esteem is to talk about your feelings with a close friend. Hold a self-esteem-building session with a friend using this worksheet.

1. Complete the following sentences:

   a. I am afraid people won't like me because

   ____

   b. I am afraid people will reject me because

   ____

   c. I feel inadequate when (list situations and reasons)

   ____

2. Read your answers to 1a through 1c out loud to a close friend. Talk with your friend about why you feel the way you do. Then have your friend help you complete the following sentences:

   a. People like me because

   ____

   ____

   b. Some of my best traits are

   ____

   ____

   c. When I'm feeling unsure of myself, I can

   ____

3. Repeat steps 1 and 2, trading places with your friend.

Chapter 9

Activity D: Using Community Resources

Finding Help When Stress Is Overwhelming

If you or someone you know is facing overwhelming stress, you can find people in your community who can help. Select an individual therapist, church, synagogue, counseling center, school, or other resource that might be able to help you in times of stress. The Yellow Pages can help you make a selection. Try to contact this individual or resource, and ask the question below (see question 1a). Use the following questions to evaluate the source you selected:

Name of institution ____

Address ____________________________________ Phone number _______________

Name of minister, priest, rabbi, spiritual leader, counselor, or therapist _______________

1. Were you able to get through to the person you wanted to talk with?

Yes _______ (go to question 1a)   No _______ (go to question 2)

   a. If yes, ask, "If someone were experiencing great stress in his or her life, what could you do to help?"

   Response: ____

   ____

   ____

   b. Did this answer seem helpful?

   Yes, because ____

   ____

   No, because ____

   ____

   c. Do you think you would feel comfortable confiding in this person?

   Yes, because ____

   ____

   No, because ____

   ____

2. If you were not able to talk to the person when you called, did the person call you back in a reasonable amount of time?

Yes _______ (go to question 1a)   No _______ (go to question 1c)

Chapter 10

Activity A: Coping

Working Through the Stages of Grief

Listed below are the five stages of grief. For each of these stages, list several activities that could help you express and work through the feelings associated with that stage.

Stage 1: Denial

Stage 2: Anger

Stage 3: Bargaining

Stage 4: Depression

Stage 5: Acceptance

Chapter 10

Activity B: Using Community Resources

Support Groups for the Grieving

Find out what support groups are available in your community for people who need help adjusting to death and dealing with their feelings.

1. Look in your telephone directory under such headings as "Crisis Intervention Services," "Human Service Organizations," or "Support Groups." The index may suggest additional headings under "Bereavement Counseling."

2. Look in your telephone directory under Hospices. The hospices may be able to direct you or may have support groups available.

3. Call local hospitals and ask if they can give you a referral.

4. Call organizations such as the American Cancer Society or a local AIDS group. They may know of local support groups for the grieving.

5. Contact local churches. They may have support groups of their own or have information about others.

6. Contact each support group you find and ask the following questions:
   - Is there a charge for the program? If so, how much?
   - Who qualifies for the program?
   - Does the program have a spiritual component?
   - Does the group have a volunteer program to provide additional support beyond group counseling?

Consider the circumstances under which you might refer a person to a specific program. For example, a religious person might prefer a support group with a religious emphasis. A person with little money would need a program or services provided free of charge. Some groups may focus on specific situations, such as the death of a child or spouse or death from a particular illness. Make a list of the different types of programs for future reference.

____

____

____

____

____

____

Chapter 10

Activity C: Communicating Effectively

Writing a Condolence Letter

A personal letter of condolence is often more comforting than a preprinted sympathy card. Sometimes such letters may even be saved in family albums and passed on to future generations. However, people hesitate to write them because they do not know what to say. A good condolence letter would do the following:

- let the bereaved persons know you understand what a crushing loss they have experienced
- speak well of the deceased
- share personal memories of the deceased, if possible
- express a sincere desire to be of help

1. Sometimes you may not know the deceased person, but you may be close to a family member and want to express sympathy. Write a letter of condolence to a close friend who has lost a loved one you have not met.

2. Imagine a good friend has died, but you have not met the family. Write a letter consoling a family member you have not met.

Chapter 10

Activity D: Communicating Effectively

Communicating With Those Who Are Grieving

Think about what is helpful or not helpful to a grieving person. Then answer the following questions:

1. You see Mercedes in the hall. She has come to school for the first time since her boyfriend was killed by a drunken driver. What should you say?

2. Ramon's girlfriend, Reiko, broke up with him three weeks ago. He is still depressed and grieving. Sheila has tried to persuade him to go out with her attractive friend, but he says he is not ready. Sheila wants you to help her persuade Ramon to go out with other girls and forget Reiko. What would you tell Sheila?

3. Your cousin's best friend has died. He does not see any point in going to the funeral because it won't do his friend any good, and he is afraid it will be depressing. He asks for your advice. What do you say?

4. Carla's cat was hit and killed by a car earlier this week. Since then she has been distant and irritable. What could you do to show your support?

Chapter 11

Activity A: Solving Problems

Peer Counseling Groups

Many schools across the country have peer counseling groups. The purpose of these groups is to train teenagers to listen to other students and to be aware of danger signals. These groups serve as a first line of defense in suicide prevention, and when a crisis arises, members of the group alert a qualified adult about the situation.

1. Does your school have a peer counseling group? If so, how many students are involved? How are the peer counselors trained? How do students with problems know who the peer counselors are? What kinds of problems do peer counselors deal with?

____

____

____

____

____

____

____

2. If your school doesn't have a peer counseling group, talk to a school counselor or someone at a local crisis center or mental health facility to find out how to start a peer counseling program. Outline a plan for starting a program.

____

____

____

____

____

____

____

Activity A: Being a Wise Consumer

Comparing Medicine Prices

Anyone who has purchased OTC drugs recently knows that they can be very expensive. Learning to be a wise consumer by comparing costs before making a purchase can help you save money.

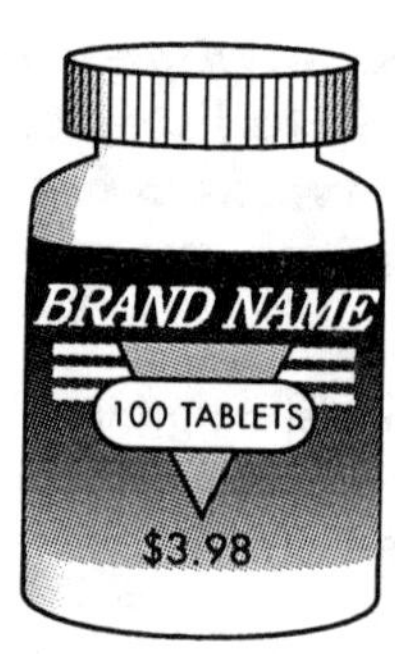

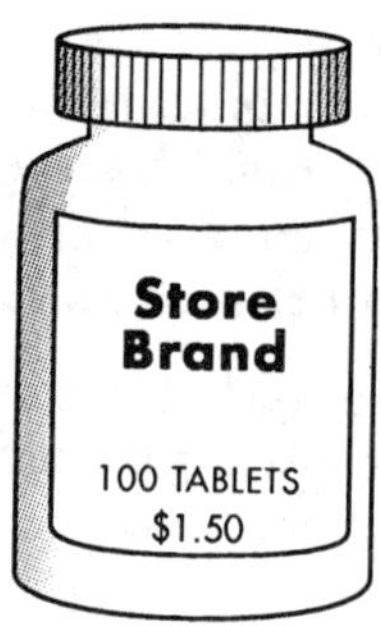

1. How would you describe the packaging of each drug?

____

____

2. Which kind of packaging has more appeal? Which is more recognizable? Explain your answers.

____

____

____

3. Why do you think brand-name manufacturers choose the type of packaging that they do?

____

4. Which drug costs less? Why do you think so?

____

____

5. How could you make sure that the contents of each brand are the same or equally effective?

____

____

Chapter 13

Alcohol: A Dangerous Drug

Activity A: Resisting Pressure

Hidden Persuasion

Advertisers often use propaganda or hidden persuasion techniques to get their audience to buy their product. Use the following activities to examine advertising techniques:

1. Search a number of magazines for advertisements featuring alcohol products. Find at least five, and examine the accompanying words and photographs. How is the alcoholic beverage photographed and in what surroundings? What types of activities are featured in the advertisements? How do the people who are drinking the beverage look?

2. Find examples of ads that attempt to associate their alcoholic beverage with the following characteristics or activities. Write the name of the product and the ways the ad makes this association. Fill in as many as you can.

   a. romance ____

   b. sex ____

   c. sophistication ____

   d. good times ____

   e. friendship ____

   f. discriminating taste _____________________________________

   g. cool refreshment ____

   h. success ____

   i. other ____

Activity B: Resisting Pressure

Resisting Pressure to Drink

Part I

Assume you are at a party where everyone else is drinking. When you refuse several drinks, the people at the party start teasing you and will not drop the subject. Assess the following explanations and add one response of your own. Rank, by number, the explanations according to their effectiveness. Number 1 would indicate the most effective and number 10 the least effective. Consider what would be most comfortable for you. Also consider your purpose. Is it to end the conversation quickly? to make your point forcefully? to set an example for your friends?

_____ My parents trust me.

_____ Alcohol makes me sick.

_____ I get terrible hangovers.

_____ I need to be in good physical shape tomorrow.

_____ I need to be clearheaded tomorrow morning.

_____ Drinking is wrong.

_____ It's against the law.

_____ Alcohol tastes terrible.

_____ I enjoy myself more when I'm alert.

_____ I don't need to explain myself to you.

Your Response: _______________________

____

Part II

How could you resist pressure in the following situations?

1. Your friend drove you to the party and then had four drinks. She says she will never speak to you again if you do not ride home with her.

____

____

____

____

____

____

2. Joaquin drove Roy to an outdoor concert. Roy brought wine. Joaquin, against his better judgment, had about 16 ounces of it. Now he realizes he cannot safely drive home. Roy insists that Joaquin drive him home. Joaquin's parents are strict, and he will be in serious trouble if he is late or if he calls his parents for a ride. If Joaquin asked you for advice, what would you say?

____

____

____

____

____

Chapter 13

Activity C: Solving Problems

Strategies for Raising Self-Esteem

For each of the following examples, suggest how the behavior described shows low self-esteem. Then write what you would suggest to the person in each example to build his or her self-esteem.

1. Roy is very shy. His acne makes him self-conscious. He also thinks he is unattractive. He is intelligent and full of interesting ideas. When he drinks, he feels more confident and engages people in conversation.

____

____

____

____

____

2. Xavier is the only one of his crowd who is not on the football team. He is wiry, strong, and fast, but he is thin and light in weight compared with the football players. At parties, he does not want to seem even more different, so he drinks to be "one of the boys."

____

____

____

____

3. Afsaneh feels ordinary. She is not a good student or athlete. She gets C's in everything, except in her art class, where she is an A student. She's heard that if she drinks, she won't care about her limitations and she will become a better artist.

____

____

____

____

Chapter 13

Activity D: Using Community Resources

Locating Resources for Breaking Additiction

Part I · · · · · · · · · · · · · · ·

Imagine that your friend or relative is addicted to alcohol and asks for your advice. Find out what resources are available in your community for people who want help to break their addiction to alcohol.

1. Look in your telephone directory under Alcohol Abuse.

2. Call toll-free numbers of national organizations such as Alcoholics Anonymous and Rational Recovery. Ask for chapters nearest you.

3. Call 1–800–662–HELP. The people at that number can provide you with a list of programs in your area.

4. Call local churches. Many of them have alcohol abuse support groups.

Part II · · · · · · · · · · · · · · ·

Call each of the organizations you find, and ask the following questions:

5. Is yours an inpatient or outpatient (residential or nonresidential) program?

6. Is there a charge for the program? If so, how much is it?

7. Who qualifies for the program?

8. Does the program have a spiritual component?

9. If the program is conducted in a hospital or under medical supervision, what treatment is provided?

10. Does the program offer an opportunity for recovering alcoholics to help others recover?

Part III · · · · · · · · · · · · · · ·

Consider the circumstances under which you might refer a person to a specific program. For example, a person whose addiction is so severe that withdrawal poses serious risk might need a medically supervised inpatient program. A religious person might prefer a support group with a religious emphasis. A person without money would need a program or service that is provided free of charge. People who are lonely or whose lives need purpose might benefit from programs in which they can continue to meet regularly and to help others. You may wish to make a note of the numbers of several types of programs for future reference.

Activity A: Communicating Effectively

Communicating With Smokers

To protect themselves from cigarette smoke, nonsmokers need to learn how to communicate effectively with smokers about maintaining a smoke-free environment. How would you respond to each of the following situations?

1. Lara was walking down the hall at school when she caught a whiff of cigarette smoke. She followed the smell and saw a younger student snuffing out a cigarette inside her locker. What would you do if you were Lara and why?

2. Juan drove up to a self-service gas station and noticed that the woman at the pump next to him was smoking as she pumped gas into her car. What would you do if you were Juan? Why?

3. Cheryl's aunt is in the hospital recovering from kidney surgery. While Cheryl is visiting, her aunt's hospital roommate lights up a cigarette. Cheryl doesn't think smoking is permitted in her aunt's hospital room, and she knows smoke makes her aunt uncomfortable. What would you do if you were Cheryl?

Activity B: Using Community Resources

Interviewing a Health-Care Professional

Look through the Yellow Pages to find health-care professionals who deal specifically with lung disease or who have contact with a variety of patients. This may be a nurse at a local hospital, a family doctor, or a lung specialist. Contact one of these professionals, and request an interview with him or her. Explain that you are completing a school assignment.

Name of health-care professional: _______________________________________

Position and place of work: _______________________________________

Conduct an interview with this person, and ask him or her the following questions:

1. How often do you have contact with people who have some kind of lung disease?

   ____

2. How many of these people smoke?

   ____

3. What lung disease do you see most frequently?

   ____

4. Are there other physical problems that these patients often have?

   ____

   ____

5. What is the worst case of lung disease caused by smoking that you have seen or heard of?

   ____

   ____

6. How many of the patients get better from their lung disease?

   ____

7. In your opinion, how does smoking impact the American health system?

   ____

   ____

Chapter 15

Activity A: Making Responsible Decisions

Marcy's Dilemma

Read the following situation, and then answer the questions:

Marcy is at her friend Naomi's house. They are watching a movie on the VCR. Naomi's parents have gone out for the evening. Shortly after her parents left, Naomi's boyfriend Jake and his friend Kevin stopped by. Marcy knew Jake well, since he had been dating Naomi for several months. She only knew Kevin by his reputation, which was that of a "druggie."

At first, they all seemed to be having a good time, and Marcy was beginning to think Kevin was okay. But, just a few minutes ago Kevin pulled out a bag of cocaine and offered to share it with them. Jake and Naomi have never tried cocaine before, but seem curious about it. Marcy doesn't want to try it, but feels pressure to go along with her friends. Marcy is faced with a problem.

1. What are some of the options that Marcy has?

____

____

____

2. What are the benefits and consequences of each option?

____

____

____

3. If Marcy decides not to use cocaine, what could she do if Naomi and Jake say yes to sharing Kevin's cocaine?

____

____

4. What could Marcy say to the others to discourage their use of cocaine?

____

____

Chapter 15

Activity B: Using Community Resources

Where to Get Help

The best way to handle drugs is to simply stay away from them. However, if you or someone you care about has a drug problem, you should know where to go for help in your community.

1. Check the Yellow Pages of your local telephone book under "Drug Abuse and Treatment" or a similar heading. List all the centers, clinics, or hospitals you find under that heading.

2. Are there any listings for local drug hotlines in your area? Sometimes these numbers are listed in the front of the telephone book along with police and fire department numbers. List these numbers below.

3. There are several self-help groups for drug abusers that are based on the philosophy of Alcoholics Anonymous. Does Narcotics Anonymous or Cocaine Anonymous have a group in your area? Write their numbers below.

4. Contact one of the facilities that offers treatment for drug abuse and ask about the kind of program offered. Ask about the goals of the program, whether special programs for teens are offered, and what the costs are. Write your information on the lines below.

Activity A: Practicing Self-Care

Caring for the Male Body

During and after puberty, it is important that you take special care of your reproductive and other body systems. Read about the self-care practices below. Then complete the lists to help you to incorporate these practices into your daily or weekly schedule.

- Whenever you participate in vigorous physical activity, you should wear an athletic supporter or other protective equipment to prevent jostling of or injury to the penis and testes.
- To prevent skin or urinary tract infections, keep the penis and the testes clean. If you are not circumcised, carefully pull back the foreskin when you bathe. Cleaning the area under the foreskin will prevent bacteria from growing and causing infection.
- It is important to identify problems before they have a chance to permanently damage your reproductive system. Describe the procedures for a testicular self-examination in the lines below so that you can have the information on hand for future reference. Also list genital disorders common to males.

____

____

____

____

- Do you know who to call if you notice any abnormalities in your reproductive organs? Look in the telephone book for numbers of clinics or hotlines that can address your questions about any symptoms you may have.

____

Things to buy:

____

____

Personal habits to develop:

____

____

Things to discuss with health-care professionals:

____

____

Activity B: Practicing Self-Care

Caring for the Female Body

During and after puberty, it is important that you take special care of your reproductive and other body systems. Read about the self-care practices below. Then complete the lists to help you incorporate these practices into your schedule.

- During menstruation, you will want to use pads or tampons to prevent the discharge from leaking onto your clothes. Tampons are tight rolls of fiber that are inserted into the vagina to absorb the menstrual flow. Pads come in different sizes. They have adhesive so they will stay in place on your underwear. Both pads and tampons should be changed every four to six hours. If you notice an unusual discharge any time during your menstrual cycle, you should consult your physician or a gynecologist.
- During your menstrual period, bathe or shower every day and carefully clean the pubic hair and vulva. This helps prevent unpleasant smells and infections.
- Some women never get menstrual cramps. Some get them every month. Many women find that exercising helps relieve the cramping. Others find that a warm bath or a nap helps them feel better. Some use nonprescription medications to relieve the pain. You may wish to talk to a doctor if you frequently have severe menstrual cramps.
- To prevent urinary tract infections, always wipe yourself from front to back after using the toilet. This helps prevent bacteria from entering the urethra.
- Review the symptoms of toxic shock syndrome. List these symptoms below for future reference. Also include the procedure you should follow in a case of toxic shock syndrome.

____

____

____

____

____

____

- It is important to identify problems before they have a chance to permanently damage your reproductive system. Describe the procedures for a breast self-examination in the lines below so that you can have the information on hand for future reference. Also list genital disorders common to females.

____

____

____

____

- Do you know who to call if you notice any abnormalities in your reproductive organs? Look in the telephone book for telephone numbers of clinics or hotlines that can address any questions you may have.

Things to buy: ____

____

____

____

Personal habits to develop:

____

____

____

____

Things to discuss with a health-care professional:

____

____

____

____

Activity A: Communicating Effectively

Nonverbal Communication

Communication involves nonverbal as well as verbal messages. Improve your ability to read nonverbal messages by completing the following activities:

Part I . . . .

Recall a recent time when someone was trying to communicate something very important. Describe the conversation, including how the person acted.

____

____

____

____

Part II . . . .

Use the checklist to mark off the nonverbal cues the person gave you during the conversation described above. You can also use this checklist to assess nonverbal cues the next time someone brings up a matter he or she feels strongly about.

Cues that the person felt strongly about the issue
- ☐ looked directly at you
- ☐ spoke loudly and clearly
- ☐ used strong gestures
- ☐ stood up straight and tall
- ☐ leaned forward
- ☐ held gaze if you asked a question

Cues that the person felt unsure about what he or she was saying
- ☐ looked down or away from you
- ☐ spoke softly or hesitated
- ☐ bit nails or exhibited other nervous gestures
- ☐ stood hunched over or shifted from foot to foot
- ☐ sat slumped in a chair
- ☐ looked away if you asked a question

If you get mixed messages (for example, if the person says he or she is sure about something but gives nonverbal cues that he or she is unsure), you may have to ask clarifying questions. Check off the question or statement below that you think would have worked the best to help you get clarifying information in the conversation you described above.

- ☐ "You seem sure. But could you tell me about it again?"
- ☐ "I'm getting mixed messages from you. If you're unsure about what you want, it's okay to tell me. I won't pressure you."
- ☐ "You seem uncomfortable. Do you want to think it over or talk about it some more?"
- ☐ Other (write in) ____

Activity B: Making Responsible Decisions

Sexual Abstinence

1. List as many reasons as you can to explain why a teenager might choose to remain or become sexually abstinent. Leave some room after each reason.

____

____

____

____

____

2. Next to each reason above, write at least one word or phrase that explains how you feel about that reason. Below are some words you can choose from, but you can also write down words you think of yourself.

independent	frustrated	bad	selfless	wise	stupid
good	glad	happy	proud	relieved	worried
scared	pressured	strong	weak	satisfied	lonely
resolved	selfish	committed	mature	intelligent	healthful

3. Which reason for abstaining from sex would make you feel best about yourself? Why?

____

____

4. Write a paragraph explaining how following through on a decision to abstain from sex during your teen years would help you build self-esteem.

____

____

____

____

____

____

____

Activity A: Setting Goals

Taking Steps to Meet Future Goals

It will not be long before you will start thinking about setting goals to meet during your adult years. For each type of life goal listed below, write down one goal you think you will want to meet during your adult years. For each goal, list a step you can take now to help you move toward that goal.

- Financial or career goal: ____
  Step I can take now to move toward that goal:

  ____

- Lifestyle goal: ____
  Step I can take now to move toward that goal:

  ____

- Interpersonal or family goal: ____
  Step I can take now to move toward that goal:

  ____

- Fitness or health goal: ____
  Step I can take now to move toward that goal:

  ____

- Intellectual goal: ____
  Step I can take now to move toward that goal:

  ____

- Spiritual goal: ____
  Step I can take now to move toward that goal:

  ____

Activity B: Coping

Caring for an Older Person

Imagine an older person you care about has suddenly become ill and no longer can care for himself or herself. Think about the decisions you might have to make about getting care for this person and your feelings about these decisions.

1. Could the older person come to live with you? Do you have room in your home?

   How would you feel about this person living with you?

2. Does the older person need nursing care? Can this be provided in your home, or does the person need to live in a nursing home?

   How do you feel about the person's need for nursing care?

3. Does the person need constant supervision? Can different people be home at different times of the day to help take care of the older person?

   How do you feel about possible restrictions on your freedom or privacy?

4. Can the older person afford the care he or she needs? Will Medicare or Medicaid help pay for the care? Will your family have to make financial sacrifices to help care for the older person?

   How do you feel about making financial sacrifices for this person?

5. The older person may have wonderful insights, stories, and memories to share with you about your family. He or she may have interests you can pursue together. How do you feel about forming a stronger relationship with this person?

Activity C: Using Community Resources

Learning About Contraception

Every person who decides to become sexually intimate should have a thorough knowledge of the contraceptive devices available to him or her. The first step when considering contraceptive methods is to discuss this decision with a parent or a guardian and a family physician. Seeking advice from a physician is especially important because he or she can offer advice concerning not only the devices themselves but where they can be obtained. Pick a contraceptive from table 18-7 on page 396 and research the following questions:

1. What is the failure rate of the contraceptive?

____

____

2. Are there dangerous side effects associated with the contraceptive?

____

____

____

3. How does the contraceptive work? Be specific.

____

____

____

4. Does the contraceptive offer protection against sexually transmitted diseases as well as pregnancy?

____

____

5. Does the contraceptive require a prescription?

____

6. How much does the contraceptive cost?

____

Chapter 19

Activity A: Using Community Resources

How Can a Counselor Help a Family?

Most communities have counseling services available to help families with problems. You might find the services listed for marriage, family, child and individual counselors in the Yellow Pages of a telephone book. To find out how these services assist families, telephone them or arrange to visit the agency in person. Ask the following questions:

1. What are some problems that your service has helped families deal with?

   Service A ____

   Service B ____

   Service C ____

2. What course of action does your service follow to help families with their problems?

   Service A ____

   Service B ____

   Service C ____

3. Besides family counselors, what other specialists do you have on staff?

   Service A ____

   Service B ____

   Service C ____

4. How much does your service cost, and how many sessions do you recommend?

   Service A ____

   Service B ____

   Service C ____

Activity A: Practicing Self-Care

Making Positive Self-Statements

To function well in life, all people, whether they come from abusive families or not, need to have self-esteem. One of the best ways of building self-esteem is through positive statements you tell yourself about your personal qualities, things you do, things you feel, and things you think. Some examples of affirmations are listed below.

- I did a good job getting my homework done today.
- I ate a healthful breakfast today.
- I played that piece of music really well.

- I did a good job stopping myself from yelling at my brother.
- I make my life interesting and fun.
- I am a good person.

As you can see, a positive self-statement can be about specific things you do on a certain day, general qualities you have, or ways you want to be treated. Use the calendar below to write down a different positive self-statement every day for the next week. Tell yourself the statement when you wake up in the morning and remind yourself of the statement at different times during the day. If you think of new statements during the day, you can write those on your calendar as well. At the end of the day, go over the statements for that day and look at the statement or statements for the next day. After a week, think about ways that making positive self-statements helped build your self-esteem and write another set of statements for the next week.

Positive Self-Statements						
Monday	Tuesday	Wednesday	Thursday	Friday	Saturday	Sunday

Activity B: Using Community Resources

Help for Abusive Families

Pick one of the resources in your community that helps members of abusive families. Choose one of those listed below or another resource you are interested in.

community health center	department of family services
substance-abuse program	parents' support group
rape-crisis hotline	child-abuse hotline
runaway hotline	center for battered women
emergency food pantry	

1. Write the name of the community resource, its address, and phone number.

____

____

2. Describe the type of service or services that you think the resource offers.

____

____

____

3. Call the community resource. Use the questions below to find out more about it. Add your own questions.

- What services does the agency offer?
- To whom are these services available?
- When are the services available?
- How much does it cost to use these services?
- What does a person have to do to get access to these services?
- What qualifications do the staff of the agency have?
- How is the agency funded?
- What should a person do in an emergency if he or she needs the kinds of services the agency offers?

4. Write a summary of your telephone inquiry. Post the summary on a classroom bulletin board.

Chapter 20

Activity C: Solving Problems

What Is Sexual Harassment?

When does sexual harassment happen? Who are the victims, and who are the abusers?

Sexual harassment is a form of abuse that occurs when a person becomes the recipient of unwelcome sexual behavior. Anyone, male or female, can be a victim or a perpetrator of sexual harassment.

Is there a legal definition of sexual harassment?

The government has defined two types of harassment. The first type, called "quid pro quo harassment," closely resembles blackmail or bribery. It occurs when a person of real or imagined authority makes sexual requirements of a student or employee in exchange for benefits or under threat of losing benefits. A supervisor who implies to an employee that he or she will get a promotion if he or she is willing to accept sexual advances is committing sexual harassment. A teacher who threatens a student with a failing grade if he or she complains about the teacher's sexual advances is committing sexual harassment.

The second type, called "environmental harassment," occurs when unwelcome sexual conduct creates a hostile or offensive environment for the employee or student. For example, a student has been the object of unwelcome sexual advances by a classmate could experience anxiety, feelings of guilt, trouble concentrating, headaches, or depression. Even though this student may have successfully rejected the advances, the situation has created a hostile environment for the student. This constitutes harassment.

The wording of such definitions varies depending on whether the definition applies to a business, school, or university.

What is "unwelcome sexual conduct"?

Unwelcome sexual conduct can be spoken, unspoken, or physical. Examples of **spoken** sexual conduct are sexual innuendoes, suggestive comments, sexually explicit jokes, suggestive whistling, graphic sexual descriptions, making remarks about a person's sexual activity, suggestive or insulting sounds, remarks of a sexual nature regarding a person's clothing or body, repeated sexual propositions, and sexual threats. Examples of **unspoken** sexual conduct are displaying sexually suggestive objects or pictures, leering, making obscene gestures, and touching oneself sexually in front of others. Examples of **physical** sexual conduct are inappropriate touching, removing another person's clothing without permission, pinching, and unnecessary brushing against the body.

Why does sexual harassment happen?

People do offensive things for different reasons. Sexual harassment is usually an attempt to show or exercise some kind of real or imagined power over another person. Forcing or coercing someone to accept or not object to unwelcome sexual conduct reinforces a false sense of authority or power. People who commit sexual harassment may or may not recognize his or her actions as harassment and in some cases may not think of his or her own conduct as unwelcome. However, just because the offender does not recognize his or her offense does not mean that harassment did not occur.

Who decides what is sexual harassment?

The government has attempted to define sexual harassment as clearly as possible. However, not everyone has the same sensitivity when it comes to sexual harassment. What is considered sexual and offensive to one person may not be sexual or offensive to another. Sensitivity may vary with cultural background, individual upbringing, or personal experience. A very important word in to remember is "unwelcome." Sexual harassment begins with behavior that is unwelcome to the person that receives it. This does not mean that saying "hello" or accidentally brushing up against someone while waiting in line constitutes sexual harassment. Even if the "hello" and the physical contact were unwelcome, neither of these actions could be considered sexual harassment. If, however, the "hello" was accompanied by a suggestive gesture or someone kept "accidentally" brushing up against the same person, either of these actions could be considered sexual harassment. The more aware people are of how they come across to other people, the less likely it is that sexual harassment will occur.

Are you being sexually harassed?

If you think you may be the victim of sexual harassment, take time to answer the following questions:

	yes	no
1. Are you insulted by the comments or behavior?	☐	☐
2. Do the comments or behavior make you angry?	☐	☐
3. Do the comments or behavior embarrass you?	☐	☐
4. Do the comments or behavior fill you with a sense of dread or guilt?	☐	☐
5. Do the comments or behavior make you feel ashamed?	☐	☐
6. Are the comments or behavior interfering with your ability to concentrate on your school work?	☐	☐

If you have answered "yes" to any of these questions, you may be the victim of sexual harassment. Know that you are not to blame for your situation and that there are things you can do to stop it. If a student, group of students, teacher, or administrator is harassing you, report the behavior to your parents, guardian, or a trusted adult at school. According to federal law, all schools must have a procedure that will resolve a harassment situation promptly and completely.

Are you committing sexual harassment?

It takes a great deal of sensitivity to recognize that a behavior you have been practicing may be hurtful and insulting to someone else. If you are wondering whether your comments or behaviors could be classified as sexual harassment, ask yourself the following questions:

	yes	no
7. Would I want my comments or behaviors to appear in the newspaper or on TV?	☐	☐
8. Is this something I would say or do if a family member or a close friend were present?	☐	☐
9. Is this something I would want someone else to say or do to a family member or a close friend?	☐	☐
10. Is this something I would say or do if the other person's family or close friends were present?	☐	☐

If you have answered "no" to any of these questions, your actions may constitute sexual harassment. If so, it is up to you to stop the behavior immediately, even if the other person does not confront you. Remember that you are responsible for your comments and behavior.

Activity D: Setting Goals

Managing Anger

Learning to manage anger is an important step toward becoming an emotion-ally mature and healthy person. But doing these things is not easy. It often helps to set intermediate, smaller goals that you can reach in a shorter period of time. Read the list of goals below, and choose one that you would like to meet.

- Recognize early signs of anger.
- Be able to walk away from situations that make you angry.
- Stop yourself from always yelling when you get angry.
- Stop yourself from hitting, kicking, or breaking something when you get angry.
- Find a positive way to channel your anger that works effectively for you.
- In a situation of conflict, be able to talk about what is making you angry in a calm, assertive way.
- In a situation of conflict, be able to talk with the people involved in the conflict in a respectful, reassuring way.
- Other __________

1. Write down the goal you are setting for yourself in your own words.

   ____

2. Give one or two reasons why you would like to meet that goal.

   ____

3. Make a list of things you can do to help you meet that goal.

   ____

4. List the way or ways you will know, after a month, if you have met your goal. Deciding how you will know if you have met your goal will help you set a more realistic goal.

   ____

5. On a separate piece of paper or in a journal, keep track of situations that make you angry. Date each incident, tell what happened, and write down how you handled your anger, focusing on your goal. Keep track of these situations for a month.

6. Write down a tangible reward you can give yourself for meeting your goal.

   ____

7. Set a new monthly goal for managing anger and for finding positive ways of channeling it.

Activity A: Being a Wise Consumer

Evaluating Advertisements

Read this imaginary advertisement for a cold remedy. Then answer the questions that follow.

1. What is this advertiser's main claim for this product? Do you think the claim is true? Why or why not?

____

____

____

2. How might this product actually help someone who has a cold?

____

____

____

3. Based on this advertisement, would you buy this product? Why or why not?

____

____

4. Where do you think a product like this would be advertised or sold? Look in such a source or a store for a similar product. Can you detect any false claims made by the advertisers of this product? Explain.

____

____

5. What can you do to help stop advertisers from making false claims about medicines? Do you feel you have a responsibility to take action? Explain.

____

____

Chapter
22

Activity A: Communicating Effectively

Preventing the Spread of STDs

People who decide to become sexually active are taking a risk with their health and the health of their partners. Partners who know each other's sexual history are better able to make responsible decisions about contraception and risk reduction. Knowing what questions to ask a partner about his or her sexual history is vital to preventing the spread of STDs.

1. Review the STDs and the risk behaviors associated with them. In the space below, list all the questions a couple should ask each other about their sexual history and risk behaviors before engaging in sexual intercourse.

____

____

____

____

____

____

____

____

____

____

____

2. What is the safest way to determine whether either sexual partner is carrying an STD? Do negative test results guarantee safety from STDs? Why or why not? Why is it extremely important that sexual partners be truthful with each other?

____

____

____

____

____

Chapter 23

Activity A: Using Community Resources

Local HIV Testing

Answer the following questions to find out about local HIV testing:

1. Where in your community could you go for an HIV test?

____

____

2. What type of counseling does this testing center offer before and after the HIV test? Are the tests confidential?

____

____

3. What test or tests does this center use?

____

____

4. How much does a test cost?

____

____

5. How soon would you get the results?

____

____

6. What are the laws in your state regarding the testing of individuals? Have these laws changed? If so, how have they changed?

____

____

Activity B: Being a Wise Consumer

Magazine Advertisements

Think of two ads you have seen recently in a magazine or on television that use a sexual theme to sell a product. Describe them, and then answer the questions that follow:

<table>
<tr><td>1st Advertisement
product:

message:</td><td>2nd Advertisement
product:

message:</td></tr>
</table>

1. Is the emphasis on the product being sold? Is the emphasis on sexual activity?

____

____

2. What behavior does each ad encourage?

____

____

3. To whom is each ad directed?

____

____

4. In your opinion, does the message being sent in either ad show irresponsible behavior? How?

____

____

5. If you feel one of the ads is irresponsible, write to the manufacturer of the product explaining why. Read your letter to the class. As a group, send the letters to the manufacturers.

6. Does irresponsibility in advertisements bother you enough to keep you from buying some products?

____

____

Activity C: Making Responsible Decisions

Teenagers and Sex

1. List as many reasons as you can to explain why two people might have sex. Leave some room after each reason.

____

____

____

____

2. Next to each reason write a word or phrase that best explains how you feel about the reason. Here are some feelings you can choose from, but you can also use your own. You may use several words to explain how you feel about each reason.

good	harmful	wrong	immature
bad	healthful	stupid	selfish
right	mature	intelligent	unselfish

3. Write what you feel are the best reasons for not having sex.

____

____

____

____

4. Considering the concerns about HIV, at what point during a relationship do you feel a couple should discuss the possible consequences of having sex? Explain your answer.

____

____

____

____

Chapter
24

Activity A: Making Responsible Decisions

Giving Good Health-Care Advice

For each situation below, give advice to the person involved. Explain you answer.

1. Karena's father died in his fifties of a heart attack. Karena is currently healthy. She does not like sports and does not exercise much. Her favorite foods are ice cream and deep fried foods. Because most of her friends smoke, Karena has started smoking, too.

2. Volker's mother has cancer, and so does his uncle. He is athletic and spends most of his time outdoors. He has just started smoking about one cigarette a day.

3. Carol is pregnant for the first time. She avoids drugs, but she occasionally likes to have a glass of wine with dinner.

4. Mr. Garcia rarely exercises and is slightly overweight. He smokes and drinks in moderation. At his last medical checkup, his blood pressure was 150/100.

5. Both of Robert's parents died of cancer while they were still young. Robert is in his early twenties and wants to know what he can do to reduce his risk of developing cancer.

6. Kevin's mother has sickle cell anemia. He does not see the point in taking care of himself because he is going to develop sickle cell anemia anyway.

Chapter 24

Activity B: Practicing Self-Care

Knowing When to See a Doctor

Which of the people below should consult a doctor immediately? What do their symptoms suggest?

1. All her life Maxine has had occasional trouble with constipation.

2. George's mole has begun to thicken and grow.

3. Colleen is suddenly thirsty all the time and constantly urinating.

4. Michael has a bad cold, and he is hoarse and coughing. He does not have a fever.

5. Rubin has had a large, raised mole for years.

6. Maria has experienced stiffness, dizziness, and vision problems.

7. Sung recently started passing thin, dark-colored stools.

8. Joan has a large sore on her knee that has not healed.

9. Mr. Hamaguchi has had a persistent cough for months. Otherwise he feels fine.

Chapter 24

Activity C: Using Community Resources

National and Local Organizations

Many of the diseases you have studied in this chapter are the focus of national and local groups. Some nonprofit corporations exist solely to study the causes and treatment of a particular noncommunicable disease. Support groups exist to help people and their families live with these diseases.

1. For each of the diseases you have studied, find the national groups that exist to study it and to help people who have the disease. Then find out if local chapters are available in your community or nearby. Look in the telephone directory under the name of the disease, call 800 information, or call hospitals to get the information you need.

____

____

____

____

____

____

2. For each of the diseases you have studied, find out if local support groups exist to help people with these diseases.

____

____

____

____

____

____

____

Activity D: Resisting Pressure

Making Healthful Lifestyle Decisions

For each situation below, propose strategies for resisting pressure to engage in behaviors at risk for cancer, heart disease, or a stroke.

1. Minetta tells Jean she is too pale and would look better with a tan. She wants Jean to spend the day lying in the sun with her. What should Jean say to Minetta?

____

____

____

2. Greta's boyfriend tells her she is too fat. She has gained weight since she quit smoking. He suggests she take up smoking again to help her lose weight.

____

____

____

3. Shane is on a sensible diet and is almost down to a healthy weight. His girlfriend has him over for dinner regularly and is hurt when he does not eat the rich desserts she prepares.

____

____

____

4. Kelly is healthy and does not have trouble maintaining her weight. Her boyfriend always wants to stop at an ice cream parlor after dates. Because ice cream is high in cholesterol, Kelly wants to avoid it. Her boyfriend says she is too young for a heart attack and does not have a weight problem. He does not understand why she wants to avoid ice cream. He feels embarrassed to order something unless she does, and the ice cream parlor serves only ice cream.

____

____

____

Activity A: Practicing Self-Care

Avoiding Tobacco Smoke

You have learned that tobacco smoke can impair health in a number of ways. In this chapter, you learned that it is a common indoor pollutant. Even a nonsmoker can suffer impaired health by breathing tobacco smoke in the environment. Think about the strategies a nonsmoker can use to avoid breathing tobacco smoke. Write your strategies on a sheet of paper.

1. Consider the smoking laws in your area. You may have laws prohibiting smoking in some public places but not in others. How can a nonsmoker avoid smoke in the following situations?

   a. At work

   ____

   b. In public places

   ____

   c. In seeking entertainment

   ____

2. Nonsmokers are often exposed to smoke at home, in social situations, or when smokers ignore prohibitions against smoking. How should a nonsmoker who wants to avoid smoke handle the following situations?

   a. A family member smokes and refuses to quit.

   ____

   b. A good friend says, "Do you mind if I smoke?"

   ____

   c. The nonsmoker has made an effort to sit in a nonsmoking area of a public place, but a person nearby lights a cigarette anyway.

   ____

Chapter 25

Activity B: Practicing Self-Care

Protection From Noise Pollution

Noise pollution is a fact of life. You have read about the ways it can damage your health. Some sources of noise are under your control. You have no control over others. Examine the sources of noise pollution in your life, and decide what you can do to protect yourself.

1. List sources of constant or frequent loud noise that you have no control over and that are not connected with a voluntary activity. Check Figure 25-13 to determine which are threats to your health. Is there any change you can make in your life that will allow you to avoid the noise, such as changing your route to school?

2. List sources of noise for which you can control the volume. Are you willing to reduce the level of noise from these sources?

3. List voluntary activities, such as rock concerts, that frequently include noise that you do not control. Are you willing to limit any of these activities? How could you protect your hearing at these events?

Activity C: Solving Problems

Caring for the Environment

How could the people in the following situations act responsibly to protect their health or their environment? What should they say to their friends? Share your responses with your classmates.

1. Tammy and Marcy are planning to see a movie together. They live in the same neighborhood, but Tammy suggests they both drive to the theater and meet at the box office.

   ____

   ____

   ____

2. A local youth group has just had a meeting at which refreshments were served and the members are cleaning up. Justine wants to separate the recyclable materials from the trash in the wastebaskets. The other members are in a hurry and want to dump the wastebaskets in the trash bin and leave.

   ____

   ____

   ____

3. Miguel and Peter are planning a party. Miguel wants to ask guests not to smoke, for the sake of the nonsmokers. Peter says that if they do that, the smokers will not come, and some of the most popular guests smoke.

   ____

   ____

   ____

4. Kerry and Betsy are housemates. Kerry wants to start composting biodegradable garbage. Betsy thinks compost heaps are disgusting and does not want rotting garbage in the backyard.

   ____

   ____

   ____

5. Isabella is on the dance committee and the dance is in progress. The music is too loud. She has it turned down, and some of the students complain.

   ____

   ____

   ____

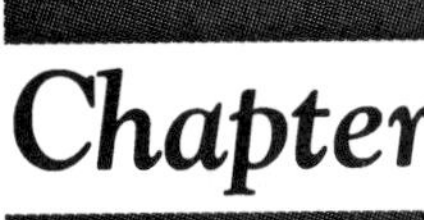

Chapter 25

Activity D: Using Community Resources

Community Organizations

Some changes require large groups of people to work together. Sometimes you can make a greater impact by lending your individual support to a larger effort. A number of private organizations exist to help protect the environment.

1. Find out what private organizations in your community work toward protecting the environment or address environmental-health issues. Include both local groups and local chapters of state, national, or international groups. The organizations listed in this chapter may be able to tell you how to contact local chapters.

2. Find out what actions each organization takes to address its issues.

3. Contact each organization and find out what types of support the group needs from individuals.

4. If you wish to contribute to environmental health through an organization, evaluate the impact of these groups. In which group would your contributions have the greatest effect?

Chapter 26

Activity A: Making Responsible Decisions

Finding a Primary-Care Physician

Use the following questions to find a primary care physician, and conduct an interview with that doctor. Based on your answers, decide whether you would want to become a patient of that doctor.

1. List the names of three primary care physicians who are highly recommended by close friends or family members.

   _______________________ _______________________ _______________________

2. For each physician named above, find out why he or she received a strong recommendation.

   a. ____

   b. ____

   c. ____

3. Based on the reasons listed above, decide which physician you want to interview.

   ____

4. What is the address and phone number of the physician?

   ____

5. What are the physician's hours? How much does he or she charge for a routine office visit?

   ____

6. How many years of experience does the physician have? _________
   When and from where did he or she graduate? ____________________________

7. Does the physician seem willing to answer your questions? _________

8. Does the physician seem concerned and knowledgeable? _________

9. Does the physician have specific times he or she is available to answer questions over the phone?
   _________ If not, is a nurse or other medical staff available to discuss medical questions with the
   physician's patients? _________ If so, when is the service available? ____________________________

10. Can you reach your physician after hours in case of an emergency? _________ If not, what should you
    do in an emergency situation? ____

11. Based on the answers to these questions, would you want to select this doctor to be your primary care
    physician? _________ Why or why not? ____________________________________

Activity B: Using Community Resources

Free or Low-Cost Health Services

What free or low-cost health services are available in your community? Look in the telephone book to find names and phone numbers of state and county health services and free clinics in your area. Call one of them for information, and record the information in the list below.

Name of clinic ____

Address of clinic ____

Phone number of clinic ____

Name and job title of person you talked to ________________________________

____

What services does the clinic offer? ______________________________________

____

Who does the clinic serve? ____

____

Are there any requirements for a person to receive services at this clinic? _________

If so, what are they? ____

____

Does the clinic charge for its services? If so what is the range of fees? ____________

____

Who administers the clinic? ____

Where does the clinic get its funding? ____________________________________

Is any written information available about the clinic? If so, can the staff send you a copy? _____________

Under what circumstances might you use the services of this clinic? _____________

____

Activity C: Resisting Pressure

Responding to People Selling Products

People selling quack health-care products often go door-to-door or make phone calls to solicit business. By contacting potential customers directly, the sellers create extra pressure for people to buy their products. Thinking about your response to salespeople can help you resist such pressure. Read each of the high-pressure scenarios below, and use the prompts to frame a verbal response.

1. It's dinnertime. You receive a phone call from a man selling a water purifier. The man claims that it removes any and all toxins—from chlorine to lead to pesticide residue—in drinking water. The water purifier is on sale for $299. To get this deal, you have to send a check today. How would you respond? Choose one of the prompts below and write out what you would say.

    "I don't need a water purifier because ____ ."

    "Before I would buy a water purifier, I would need written information about ____________

    ____ ."

    "Your water purifier might be useless because ________________________________ ."

    "I don't buy products over the phone unless __________________________________ ."

    "You interrupted my dinner, so ____ ."

    [Other response of your choice] ____ .

2. A friend has gone on a diet based on a low-calorie drink that he ordered from a mail-order company. He has been so happy with the diet drink that he has become a sales representative for the product, selling it door-to-door and to his friends at school. He has given you a brochure and asked if you would like to buy a case of the diet drink. How would you respond? Choose one of the prompts below and write what you would say.

    "I don't want a case of diet drink because ____________________________________ ."

    "Before I would buy a case of diet drink, I would need information on ________________ ."

    "Your diet drink could harm my health because ________________________________ ."

    "I feel uncomfortable buying such a product from a friend because __________________ ."

    [Other response of your choice] ____ .

Chapter 27

Activity A: Being a Wise Consumer

Looking at Fire Extinguishers

Several thousand people die each year in home fires. Many deaths could be prevented by following safety measures to prevent fires from starting or growing. Fire extinguishers can be used to put out a small fire before it gets out of control. It is recommended that each home have a fire extinguisher in the kitchen and one in the basement. It is also a good idea to carry a fire extinguisher in your car.

Find out about the types of portable fire extinguishers that are available. You might begin by checking at your local hardware store.

1. What different brands and types of fire extinguishers are available?

2. What is the cost of each?

3. What types of fire will each extinguisher put out?

4. Which extinguisher would be best to store in your kitchen? your basement? your car?

Activity A: Solving Problems

Responding to an Emergency

1. Sometimes we can be of most help in an emergency situation by protecting ourselves first. For example, when an airplane loses cabin pressure, adult passengers are instructed to put on their own oxygen masks first before attending to any child. What do you think are the reasons behind that rule?

2. In the space below, list several situations in which you think it would be best to protect yourself first rather than to immediately begin giving treatment. For each example, think of at least one way that you could help besides giving direct treatment.

Chapter
28

Activity B: Practicing Self-Care

First-Aid Supplies

Many of the treatments in this chapter require the use of specific equipment or materials. In case of an emergency, you may want to gather some first-aid supplies and keep them handy in your home and perhaps in your vehicle. Think about what a first-aid kit should contain for each of the following emergencies. List the items under the appropriate emergencies.

1. Wounds

____

____

2. A fracture

____

____

3. To avoid contact with an injured person's blood

____

____

4. For checking for possible head injury

____

____

5. For treatment of burns

____

____

6. To use for pressure bandages

____

____

Activity C: Using Community Resources

Emergency Medical Help

First aid is the emergency care given to a victim until medical help is obtained. For all serious emergencies, summoning medical help, taking a person to an emergency room, or calling medical personnel for advice is an important step in the treatment. In an emergency, you can lose time trying to find out what number to call or where you can take someone for care. Prepare the following information in case you ever need it:

1. Write the local emergency telephone number. ________________________

2. Write the telephone number of the nearest poison control center. ________________________

3. List the names, telephone numbers, and locations of the 24-hour medical facilities in your area.

____

____

____

4. List the names, telephone numbers, and locations of the hospitals with emergency rooms in your area.

____

____

____

5. Place a check beside each hospital you listed for item 4 that has a trauma unit.

6. Find out if there are other facilities that specialize in care for specific emergency conditions. For example, some hospitals specialize in treating children. List them here.

____

____

____

7. Write this information on a small card to keep in your wallet. You may want to make another card to post near your telephone at home if you do not already have one there.

Activity D: Making Responsible Decisions

Taking Appropriate Action

What advice would you offer to the following people?

1. You have arrived at the scene of a car accident behind another person. Several people are lying injured on the ground, in no danger of further injury. The person ahead of you asks you to help him move the victims to a nearby house to treat them.

2. Your friend bumps against a nail, leaving a deep puncture wound in her leg. She says she does not need to stop what she is doing to treat it because it isn't bleeding very much.

3. In a restroom, you see a small child cleaning an abrasion on his knee. He has rinsed it thoroughly and removed the large pieces of dirt. He is crying because trying to remove the tiny pieces hurts.

4. Near the scene of an explosion, you see a victim with a whitish-colored burn. The victim says that it doesn't hurt. A bystander is about to apply ice to the burn.

5. A cyclist without a helmet has had a bad spill and has a cut on her forehead. A bystander is attempting to treat her for shock. He is about to elevate her legs.

6. You are hiking with a friend who has been bitten by a snake that you were not able to identify. She wants to run down the trail to get medical help as soon as possible.

7. Your neighbor has injured his ankle. His wife does not know whether it is a sprain or a fracture, so she does not know what kind of emergency care to provide.

8. A family member comes home from a long walk in the snow with frostbitten toes. She is about to immerse her foot in steaming hot water.

9. Your friend has been playing volleyball on a very hot day. As he comes off the court, his face is very red. His skin is dry and hot to the touch. He says he will be fine if he can just rest in the shade.

Introduction to Health and Wellness

Evaluating Media Messages

What Is a Media Message?

The steps in creating an advertisement for a product include

1. Identifing the **target market** or **audience**—those who are most likely to buy the product being promoted. Factors that help determine the target market include age, sex, occupation, and income.
2. Selecting the **medium** or **media** to be used—these include newspapers, radio, television, direct mail, magazines, outdoor signs, window displays, and posters.
3. Determining the **basic appeal**—the way the product will be presented. A factual appeal describes what a product is, how it works, and how it is made. An emotional appeal stresses how the product will provide personal satisfaction. Sometimes a combination of the two approaches is used.
4. Choosing the **technique** or techniques for presenting the message—these include attention-getting headlines, slogans, testimonials (endorsements from celebrities or other "real people"), product characters (fictional people or cartoon animals or characters), comparison of products, and repetition.

Select five of the media listed in step 2. Find an example of an advertisement for a health-related product in each medium. Complete the chart below for each advertisement. A sample answer is included to help you get started.

Product	Audience	Basic Appeal	Medium or Media	Techniques
non-aspirin pain reliever	senior citizens	factual	magazine	attention-getting headline in large letters: "Arthritis Pain Relief"

As a class, compare the charts. Do the messages seem to change according to the target audience? How does analyzing target audience, basic appeal, and technique help you identify and evaluate media messages?

Unit 2

Health and Your Body

Evaluating Media Messages

How "In" Is Thin?

In this unit, you learned about the benefits of exercise and proper nutrition. You also learned that media messages often equate thinness with health. Find five advertisements that emphasize weight loss or reinforce the idea that thin bodies are healthy and desirable. Then answer the following questions:

1. For each advertisement, explain why you think it emphasizes the idea that being thin is healthy and desirable.

2. In the space below, briefly summarize the product advertised, the audience targeted, the basic appeal, the medium, and the techniques used in each advertisement.

3. Are there any audiences that seem to be targeted more than others? If so, why do you think this is so? Are there any other patterns that you can identify?

4. Are there advertisements that seem to make realistic claims? If so, how are they realistic?

____

____

____

____

____

5. Are there advertisements that seem to make unrealistic claims? If so, how are they unrealistic?

____

____

____

____

____

6. How might you find out if the claims are true?

____

____

____

____

____

7. Change the wording in one of the advertisements to make it more realistic or less sensational. Compare your work with that of other classmates.

____

____

____

____

Unit 4

Protecting Your Health in a Drug Society

Evaluating Media Messages

How Bad Can It Be If Athletes Do It?

Advertisers of potentially harmful products sometimes use positive images to make their product seem more acceptable. These images can make it appear as though the product is less harmful than it really is. Find three advertisements that you think use such positive images. In the chart below, identify the positive image used and the implied message. Then write your own counter-message that responds to the implied message. In the last column, write an idea for a negative image that would support your message. A sample answer is included to help you get started.

Product	Positive Image	Implied Message	Counter-Message	Negative Image
Samson's Chewing Tobacco	image of a baseball player chewing tobacco	How bad can it be if athletes do it?	Actually, it can be very harmful, causing cancer in the mouth and lips that can lead to death.	chart showing statistics of cancer deaths related to the use of chewing tobacco

Family Life, Sexuality, and Social Health

Evaluating Media Messages

Violence and the Media

Many young children watch cartoons on television. Watch at least one hour of cartoon programs yourself. Keep a record of the length of each cartoon and the number of violent images shown in each cartoon. Then answer the following questions:

1. For each cartoon program you watched, calculate the number of violent images shown each minute.

   ____

   ____

2. Do you think these programs use violence to gain and keep the attention of children? Explain your answer.

   ____

   ____

3. Do you think any of these programs promote violence as a means of solving problems? If so, how?

   ____

   ____

4. Do any of these programs promote peaceful solutions to problems? Explain your answer.

   ____

   ____

   ____

Based on your analysis of and reaction to the children's programs you watched, write a persuasive letter to a television station manager. Remember that a persuasive letter is meant to influence the reader to agree with your opinion and to act accordingly. Write the final draft of your letter on a separate sheet of paper.

Ethical Issues in Health

ISSUE: *Should terminally ill people have the right to determine how or when to die?*

Review the issues of assisted death and the living will. Enter reasons for and against the issue in the boxes on the scale below. Rank how you feel about each reason as follows: 10—most important, 5—average importance, and 2—least important. Add the ranks in each box. The total "weight" of the reasons in each box, rather than the number of reasons, should determine which direction you lean on this issue.

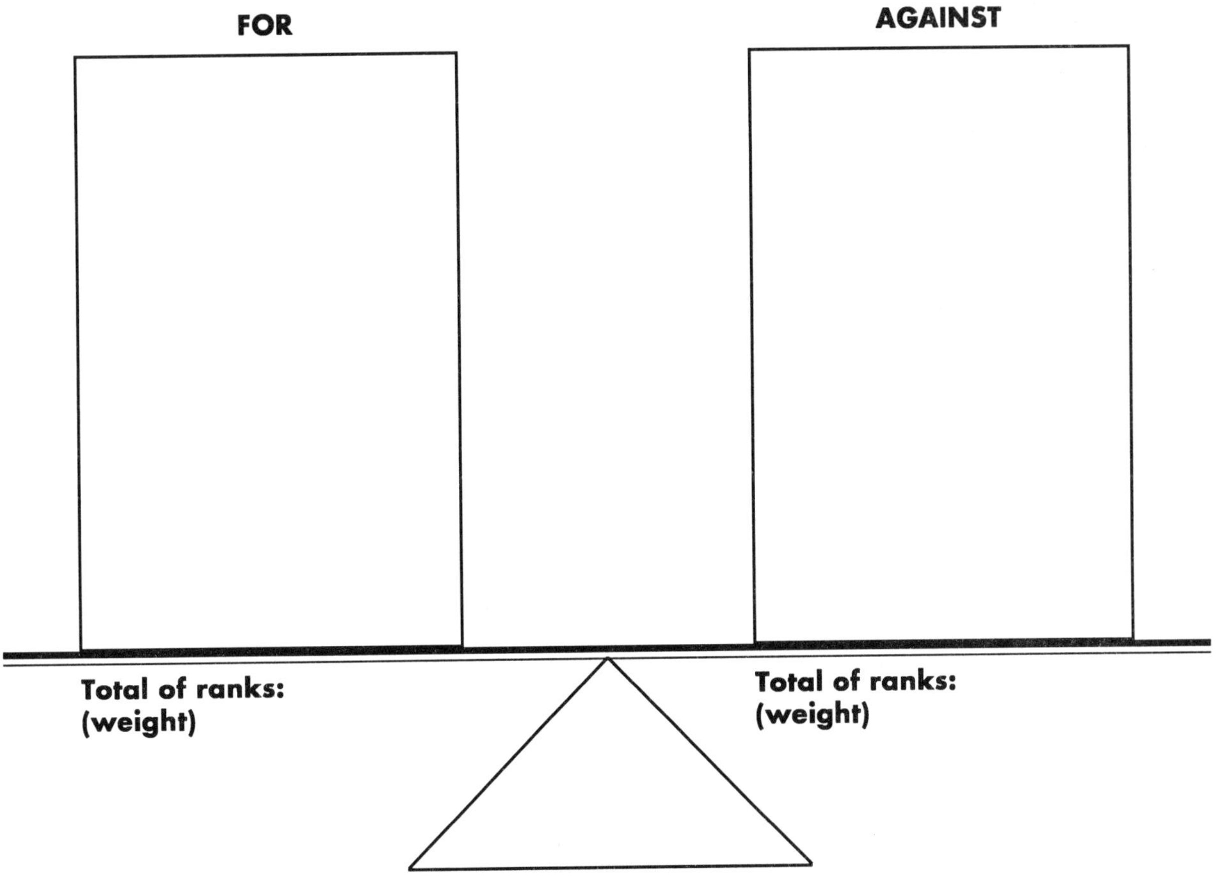

As a class, use your completed worksheets as a basis for preparing a debate on this issue.

Unit 6

Diseases and Disorders
Ethical Issues in Health

ISSUE: Should clean needles be distributed to intravenous drug users in order to slow the spread of AIDS?

Suppose that the legislature in your state is considering whether to pass a law authorizing a needle-exchange program. A group of students who share your opinion on the issue have formed an organization. You have been asked to testify before a legislative committee on the organization's behalf.

First prepare notes for your speech. What do you think the legislators need to know? What can you say to persuade them of your point of view? Then write a rough draft of your testimony. Revise your writing for clarity and style. Write the final draft of your speech below. Use an additional sheet of paper if necessary.

____

____

____

____

____

____

____

____

____

____

____

____

Students should take turns reading their speeches. After each speech, classmates should ask questions, just as legislators would do.

Unit 7

Ethical Issues in Health

ISSUE: *Should family leave be an employment benefit in the same way as sick leave is?*

Suppose that you had an opportunity to interview one of the senators from your state on the issue of family leave. Prepare a set of interview questions, and write them below. Remember that to obtain information, the interviewer must be thoroughly prepared. Before you write your questions, review the information in the ethics feature and other information you may have gained through research and class work.

____

____

____

____

____

____

____

____

____

____

____

____

____

____

____

____

When you have completed your questions, ask a classmate to play the part of the senator. Role-play the interview for the class.

Unit 8

Safety and Emergency Care

Ethical Issues in Health

ISSUE: Should gun-control laws be passed to reduce the number of deaths from gunshot wounds?

Using what you have learned about media messages, create a 30-second TV commercial that promotes your opinions about gun-control laws. You can use a direct appeal in which you try to persuade the viewers of your point of view, or you can create a narrative that tells a story dramatically.

In the space below, summarize the script and describe the visual images that will accompany it.

____

____

____

____

Use the table below to identify the audience targeted, the basic appeal, the medium, and the techniques to be used in your advertisement. Share your work with the class.

Message	Audience	Basic Appeal	Medium or Media	Techniques

Answer Key
for Life Skills Worksheets

Chapter 1

Activity A

Answers will vary. Some students may not be able to complete all sections of this form. Encourage students to complete as much of the form as they can.

Chapter 2

Activity A

1. Accept all reasonable answers. Answers will vary but should include the benefits and consequences of each option. For example, declining Jack's offer and waiting to use the phone could result in embarrassment. It could also cause the parents to worry temporarily. However, these actions prevent the possibility of serious injury from a motorcycle accident.
2. Answers will vary depending on students' values.
3. Answers will vary but should include positive consequences that could result from the decision made.
4. Answers will vary but should include negative consequences that could result from the decision made.
5. Answers will vary.

Chapter 3

Activity A

This activity is designed to enable students to practice setting realistic goals and plan how to attain them. Answers to all parts of this activity will vary depending on students' abilities and goals. Students should answer every question as thoroughly as possible.

Activity B

1. Answers will vary. Students should keep accurate and thorough records for each of the seven days. Each student should also keep a journal, but the contents of the journal should not be graded.
2. Answers will vary. Students may notice that eating within three hours of bedtime and experiencing high stress can make it more difficult to fall asleep. Students may also notice that getting eight or more hours of sleep each night improves energy level, mood, and mental alertness. If students do not notice any patterns, you may wish to help them reevaluate their records.
3. Answers will vary depending on students' current sleep schedules. Students should suggest ways to improve their sleep.

Chapter 4

Activity A

Answers will vary but students should include the ingredients for at least five products that have more than one form of sugar, fat, or sodium.

Activity B

This activity is designed to enable students to apply their understanding of nutritional requirements by creating a well-balanced menu. Answers will vary but menus should adhere to the food pyramid and satisfy the FDA's recommended daily intakes.

Activity C

1. Answers will vary depending on the products chosen. Most students will probably find that instant and cold cereals cost more than whole-grain cooked cereals.
2. Answers will vary depending on the products chosen. Most students will probably find that raw servings of whole grains have the most protein, fiber, and carbohydrates. Regular cooked hot cereals tend to have more protein, fiber, and carbohydrates than instant hot cereals. Cold cereals tend to have the least amount of protein, carbohydrates, and fiber, but tend to provide the most vitamin A and riboflavin. You may wish to explain to students that cold cereals are often fortified with vitamins and minerals. Fat content varies

dramatically depending on the individual product.

Activity D

1. Answers will vary.
2. Answers will vary but should demonstrate an understanding of the dietary guidelines presented in this chapter.
3. Answers will vary but should be thorough and should demonstrate an understanding of how to use the RDI.
4. Answers will vary.

Chapter 5

Activity A

This activity is designed to enable students to compare the characteristics of three local weight-loss programs. Students should obtain thorough information for each category. You may wish to have students prepare additional questions or present the information to the class.

Chapter 6

Activity A

1. Answers will vary. Students should recognize that a sunscreen with an SPF factor of 8 is not strong enough to protect the skin from the sun for an entire day. Students may suggest wearing additional clothing, spending time under an umbrella or another source of shade, or attempting to purchase sunscreen lotion at a local store.
2. Answers will vary but should include two reasonable solutions. For example, students might suggest looking for an alternate hiking trail or wearing protective clothing.
3. Answers will vary but should suggest that sharing combs or brushes should be avoided because such behaviors transmit head lice and ringworm.
4. By keeping a lending comb, Consuela avoids getting head lice or ringworm. However, if more than one of Consuela's friends uses the lending comb, head lice or ringworm could be transmitted.

Activity B

Answers will vary but should demonstrate thorough research.

Chapter 7

Activity A

1–3. Answers will vary but should demonstrate an understanding of how taking part in community activities promotes mental and emotional health.

Activity B

1. Answers will vary but should demonstrate an understanding of defense mechanisms and how they can preserve self-esteem. Examples of positive uses of defense mechanisms are daydreaming to gain control over a fear of flying, sublimating anger by playing piano, and compensating for feelings of failure by volunteering for a non-profit organization.
2. Answers will vary but should demonstrate an understanding of how defense mechanisms can hinder emotional growth and wellness. Examples include using regressive behaviors (such as whining) to get attention, becoming nauseated whenever conflict arises, and projecting one's own low self-esteem on others by making fun of them.

Activity C

Answers will vary. This activity is designed to help students use Maslow's Hierarchy of Needs to identify and address the individual factors that affect a mood. Adolescence is an especially difficult time because so many changes are happening at once. Adolescent minds and bodies change so quickly that everyday problems often become very complex. The physical and emotional changes are often so entwined that it is very difficult for a teenager to identify why they feel the way they do. For example, several emotional, hormonal, and physical factors could be influencing one feeling of general frustration. As a result, most teenagers spend a great deal of time feeling confused about problems that they can neither identify nor solve. Maslow's Hierarchy can be very useful because it offers a framework for classifying and ranking needs. You may wish to ask students why it is important to address the most basic needs first. *(This is because basic needs prevent other needs from being met. For example, the most talented counselor cannot help a patient if the patient is fainting with hunger. Once the basic needs are met, higher level needs can be addressed.)*

Activity D

1–6. This activity is designed to enable students to evaluate the counseling programs in their area. Answers will vary but should provide thorough and accurate information.

Chapter 8

Activity A

Answers will vary depending on students' values. Most students will probably suggest that examples 3 and 6 are inappropriate forms of disclosure. Students should write how they would respond in each situation.

Activity B

While there are no correct answers to the questions in this activity, students' answers should reflect thorough research into community resources.

Activity C

1. Answers will vary. Students may suggest that Samantha would be preserving her self-esteem by not having sex with her boyfriend. Students may also suggest ways that Samantha could talk to her boyfriend about her feelings.
2. Answers will vary. Students may suggest that although dating Angel might give Sue a superficial sense of self-esteem, his treatment of her could actually be harmful to her self-esteem. Raol appears to be more supportive.
3. Answers will vary. Students may recognize that Francine likes showing her work to Jerry because it gives her a false sense of self-esteem. In the long-run, receiving honest criticism from Arsenio will allow Francine to improve her work and will help her build true self-esteem.

Activity D

1. c
2. b
3. a, b
4. a, b
5. a
6. b
7. a, b
8. c
9. b
10. a
11. c
12. b
13. c

Chapter 9

Activity A

This activity is designed to help students apply what they know about reducing stress to their own lives. Students' diagrams will vary depending on personal preferences.

Activity B

1–3. This activity is designed to help students analyze and evaluate the success of their communication with another person. Answers will vary depending on individual experience and opinion.

Activity C

This activity is designed to facilitate positive self-disclosure. There may be some students who are not able to speak to a close friend. To avoid calling attention to these students, you may wish to give students the option of completing the assignment with a sibling, parent, or trusted adult.

Activity D

1–2. Answers will vary but should provide thorough information about the chosen resource.

Chapter 10

Activity A

Answers will vary but should include constructive behaviors that express the feelings associated with each stage. For example, one could work through denial by attending a funeral or talking to others who share in the loss. Anger could be expressed in a letter to oneself or vented though vigorous physical exercise. To work through the bargaining stage, expressing one's feelings in a journal or to a counselor could help keep these feelings in perspective. Crying or meditating could help express and get rid of feelings of depression. Acceptance could be cemented by activities that symbolize closure, such as constructing a photo album or writing a story to commemorate the lost item, person, or relationship.

Activity B

While there are no correct answers to the questions in this activity, students' answers should reflect thorough research into community resources.

Activity C

This activity is designed to enable students to apply what they know about the grieving process to a personal situation. The contents of the condolence letters will vary but should roughly follow the guidelines for a good condolence letter.

Activity D

1. Answers will vary. Students may suggest offering to listen or asking Mercedes what she needs in the way of support.
2. Answers will vary. Students may suggest that Sheila is not helping Ramon by pushing him to stop grieving before he is ready.
3. Answers will vary. Students may suggest that a funeral can assist the grieving process by helping mourners acknowledge their loss and by allowing mourning friends and family members to share their feelings and support.
4. Answers will vary. Students may suggest offering to listen or asking Carla what she needs in the way of support.

Chapter 11

Activity A

While there are no correct answers to the questions in this activity, students' answers should reflect thorough research into the resources available at school.

Chapter 12

Activity A

1. The packaging for the brand name drug is much more stylized and flashy than the packing on the generic brand.
2. Students will probably suggest that the brand-name package has more appeal and is easier

to recognize because the design is eye-catching and distinctive.
3. Students may suggest that manufacturers choose stylized packaging in order to make them attention-getting and recognizable.
4. The store brand costs $2.48 less than the brand-name brand. Students may suggest that fancy packaging costs more or that people are willing to pay more for name-recognition. Some students may observe that the amount of drug in each tablet is not available on either container. The difference in cost may reflect a difference in the size or potency of a tablet.
5. Students may suggest reading the ingredients or asking a pharmacist.

Chapter 13

Activity A

1–2. This worksheet is designed to enable students to evaluate the persuasive techniques used to advertise alcohol. Answers will vary but should be thorough.

Activity B

Part II

1. Answers will vary but should suggest a way to avoid riding in the car. Students may also suggest ways to secure a safe ride home for the friend.
2. Answers will vary but should suggest a way for Joaquin to avoid driving or riding with someone who has been drinking.

Activity C

1–3. Answers will vary but should indicate an understanding of how using alcohol can create a false sense of self-esteem. Students should also suggest alternative behaviors that build self-esteem.

Activity D

While there are no correct answers to the questions in this activity, students' answers should reflect thorough research into community resources.

Chapter 14

Activity A

Although there are no correct answers for the

questions in this activity, students' answers should
be thorough and should include reasons.

Activity B

While there are no correct answers to the questions
in this activity, students' answers should reflect
thorough research into community resources.

Chapter 15

Activity A

1. Answers will vary. Some of Marcy's options
   are trying cocaine, making an excuse to leave,
   and openly resisting pressure by telling her
   friends that she does not want to try cocaine.
2. Answers will vary but should be thorough and
   logical. Possible consequences of Marcy's
   trying cocaine include lowering her self-
   esteem, becoming addicted, being caught,
   having a bad reaction, and overdosing.
   Benefits of trying cocaine might include
   satisfying a curiosity or feeling accepted by
   her friends.
3. Answers will vary. Students may suggest that
   Marcy leave Naomi's house or ask her friends
   not to try cocaine.
4. Answers will vary. Students should demonstrate an understanding of verbal resistance.
   For example, students may suggest that Marcy
   tell the others why she feels that the risks
   associated with cocaine outweigh the possible
   benefits.

Activity B

While there are no correct answers to the questions
in this activity, students' answers should reflect
thorough research into community resources.

Chapter 16

Activity A

Students' worksheets are for personal use and
should not be graded. However, you may wish to
provide students with the following information:
- The procedures for testicular self-examination
  are the following: 1) stand in front of a mirror
  and look for any swelling of the scrotum;
  2) check for any lumps, enlargements, or
  tenderness, or changes in texture by rolling
  each testicle gently between the thumb and

fingers; and 3) notify your doctor of any
abnormalities.
- Genital disorders that occur in males include
  testicular torsion, undescended testes, inguinal
  hernia, infertility, enlarged prostate gland,
  prostate cancer, and testicular cancer.

Activity B

Students' worksheets are for personal use and
should not be graded. However, you may wish to
provide students with the following information:
- Symptoms of toxic shock syndrome are a high
  fever, nausea, diarrhea, dizziness, and a
  sunburnlike rash. A woman who has these
  signs and symptoms should remove the
  tampon and then go to a hospital emergency
  room immediately.
- The procedures for breast self-examination are
  the following: 1) stand in front of a mirror and
  look for anything unusual, such as discharge,
  puckering, dimpling, or scaling of the skin; 2)
  clasp the hands behind the head, press
  forward, and look for any changes since the
  last exam; 3) press the hands firmly on the
  hips, bow slightly, pull the shoulders and
  elbows forward, and look for any changes
  since the last exam; 4) raising one arm,
  examine the breast and armpit under that arm
  by pressing in circles with the flat of the
  fingers, feeling for any unusual lumps or
  changes since the last exam, and then
  repeat the procedure with the other breast;
  5) squeeze each nipple and look for discharge; and 6) lie down, put a pillow under
  one arm, and repeat the procedure in step 4
  while lying down.
- Genital disorders that occur in females include
  menstrual cramps, premenstrual syndrome,
  vaginitis, toxic shock syndrome, ovarian cysts,
  cancer, and infertility.

Chapter 17

Activity A

While there are no correct answers to the questions
in this activity, students' answers should reflect
thorough observations of nonverbal
communication.

Activity B

1. Accept all reasonable answers. Students'
   answers will vary but should represent an

awareness of the issues involved in adolescent relationships.

2. There are no correct answers to this question.

3. There are no correct answers to this question.

4. While there are no correct answers to this question, students should demonstrate an understanding of behaviors that build self-esteem.

Chapter 18

Activity A

Answers will vary according to students' individual goals. While there are no correct answers to this activity, students should write steps they can take now to move toward each of their goals.

Activity B

There are no correct answers to this activity.

Activity C

While there are no correct answers to the questions in this activity, students' answers should reflect thorough and accurate research of community resources.

Chapter 19

Activity A

While there are no correct answers to the questions in this activity, students' answers should reflect thorough research of community resources.

Chapter 20

Activity A

There are no correct answers to this activity.

Activity B

While there are no correct answers to the questions in this activity, students' answers should reflect thorough research of community resources.

Activity C

There are no correct answers to this activity. Students' worksheets are for personal use and should not be graded. However, you may wish to encourage students to talk with their parents, a school counselor, or another adult if they think they have witnessed, experienced, or committed sexual harassment. Students interested in finding the exact legal definitions of harassment can contact the Office of Civil Rights in the Department of Education by phone or on the Internet.

Activity D

While there are no correct answers to the questions and assignments in this activity, students should thoroughly complete all parts of the activity.

Chapter 21

Activity A

1. The advertiser's main claim for this product is the ability to cure a cold. Students' opinions about the truth of this claim may vary. Students may correctly recall that according to current research, there is no known cure for the common cold.

2. Answers will vary, but students should suggest that the product might alleviate some of the symptoms of a cold.

3. Answers will vary according to students' opinions.

4. Answers will vary. Students may correctly observe that although there is some evidence that vitamin C helps lessons the symptoms of a cold, it has not been proven to cure a cold.

5. Answers will vary according to students' opinions. Students may suggest writing a letter of complaint to the company or contacting an agency such as the Federal Drug Administration or the Better Business Bureau.

Chapter 22

Activity A

1. Answers may vary but should demonstrate a thorough understanding of the ways that STDs can be transmitted. Below is a thorough list of questions a couple would need to ask before engaging in sexual intercourse.

- Have you ever shared items such as needles, razors, or eating utensils?
- Have you ever come into contact with another person's open sores?
- Have you ever taken part in any form of oral, anal, or vaginal intercourse?

- If you have taken part in any form of sexual intercourse, were any of these interactions not protected by the correct use of a latex condom or dental dam?
- If you have taken part in any form of sexual intercourse, had any of your partners had any form of oral, anal, or vaginal intercourse, shared items such as needles, razors, or eating utensils, or come into contact with another person's open sores?

Present students with this list so that they can fill in questions that they do not have written down. Students may be overwhelmed by how difficult it is to protect oneself from STDs. You may wish to inform students that the best way to answer these questions is for both partners to be tested for STDs before engaging in any form of sexual intimacy.

2. The safest way to determine if either partner is carrying a STD is to be tested. However, even good test results do not guarantee that a STD will not be transmitted. If one partner takes part in any risky behavior after the test is taken, both partners are put at risk. It is absolutely essential to be able to trust that a potential sexual partner is telling the truth about their sexual history and will continue to tell the truth about any behaviors that could transmit a STD. However, even if partners are honest with each other, one or both partners may not remember some of their past risky behaviors.

Chapter 23

Activity A

While there are no correct answers to the questions in this activity, students' answers should reflect thorough research of community resources.

Activity B

While there are no correct answers to the questions in this activity, students' answers should reflect thorough observations and comparisons of the chosen advertisements.

Activity C

There are no correct answers to the questions in this activity. Answers will vary depending on students' opinions.

Chapter 24

Activity A

1. Answers will vary but should indicate an understanding of the risk factors for heart disease. For example, students may correctly suggest that the history of heart disease in Karena's family may put her at a higher risk of heart disease. To reduce these risks, Karena should moderate her intake of high-fat foods such as ice cream and fried foods as well as exercise regularly. Karena should not smoke because in addition to causing many other diseases smoking greatly contributes to heart disease.

2. The history of cancer in Volker's family may increase his risk of developing cancer. To reduce these risks Volker should wear sunblock when he is outdoors and should quit smoking completely.

3. Even in very small amounts, alcohol consumption by a pregnant woman can cause fetal alcohol syndrome in her baby. To protect her child from FAS, Carol should completely abstain from drinking any alcohol while she is pregnant.

4. Mr. Garcia's blood pressure is high, indicating a higher risk of heart attack. To lower his blood pressure and prevent any further atherosclerosis, Mr. Garcia should reduce his cholesterol, get regular moderate exercise, and quit smoking.

5. Robert should avoid all tobacco products and second-hand smoke, moderate or avoid drinking alcohol, get regular exercise, wear sunblock when necessary, eat plenty of grains, fruits, and vegetables, limit consumption of high-fats foods, get plenty of fiber, and avoid exposure to radiation and toxic chemicals. Robert should also get regular checkups and follow a regular schedule of self-examination.

6. Answers will vary but should indicate that Kevin cannot get the disease unless his father is a carrier of the sickle-cell trait. If students are confused by this idea, you may wish to explain the concept in terms of dominant and recessive genes. Because the sickle-cell gene is recessive, a person must be double-recessive (have two of these genes) to develop the disease. Kevin's mother has two recessive genes for sickle cell. If Kevin's father does not carry the trait, Kevin will carry the trait but will not develop sickle-cell anemia. If Kevin's father has one sickle-cell gene, the father will be a carrier but will not have the disease and

Kevin will have a 50 percent chance of developing the disease. If Kevin's father does have the disease, Kevin will also have sickle-cell anemia.

Activity B

1. Constipation is not listed as a symptom of any of the diseases in this chapter. Maxine should go to a doctor if the constipation is prolonged, causes her discomfort, or is unusual in some way. Maxine may not be getting enough fiber.

2. George should go to the doctor. A changing mole is one of the warning signs of cancer.

3. Coleen should go to the doctor. Her symptoms indicate a possibility of type 1 diabetes.

4. Hoarseness and coughing are normal when they are temporary and are associated with a cold. However, Michael should go to the doctor if the symptoms worsen, worry him in some way, or do not subside.

5. Raised moles are very common and normally do not indicate a problem unless they begin to change in some way. Rubin should watch for changes in the size or shape of his mole.

6. Maria's symptoms indicate the possibility of multiple sclerosis. She should go to the doctor.

7. A change in bowel habits is one of the warning signs of cancer. Sung should go to the doctor.

8. A sore that does not heal is one of the warning signs of cancer and also may indicate a possibility of type 1 diabetes. Joan should go to the doctor.

9. A persistent cough is one of the warning signs of cancer. Mr. Hamaguchi should go to the doctor.

Activity C

While there are no correct answers to the questions in this activity, students' answers should reflect thorough research of community resources.

Activity D

1. Answers will vary but should demonstrate an understanding that overexposure to the sun can cause cancer. Students may also correctly observe that if Jean has a pale complexion, she will be more susceptible to the damage caused by ultraviolet radiation.

2. Answers will vary but should indicate an understanding that smoking greatly increases a person's risk of having cancer, heart disease, or

a stroke. If Greta is indeed overweight, she would benefit much more from getting regular exercise and limiting her intake of high-fat foods. Students may also suggest that Greta's boyfriend's comments are insensitive and do not appear to support her desire to improve her health. Greta could talk to her boyfriend about how his comments make her feel.

3. Answers will vary but should demonstrate an understanding that regularly eating rich desserts would not help a person lose weight and could also lead to high cholesterol. Students may suggest that Shane talk to his girlfriend about how the pressure to eat rich desserts effects his attempt to improve his health. Students may suggest a compromise, such as Shane bringing low-fat desserts to dinner, or agreeing to have rich desserts only once a week.

4. Answers will vary but should indicate an understanding that even people with a normal weight can develop heart disease from high cholesterol. Students may suggest that Kurt share his concerns with his girlfriend and possibly work out another alternative. For example, Kurt and his girlfriend could go to an ice cream store that also serves lowfat ice cream, frozen yogurt, or sherbet.

Chapter 25

Activity A

1. Answers will vary depending on local smoking laws. Students should offer reasonable ways for nonsmokers to avoid inhaling smoke.

2. Answers will vary. Students should demonstrate an ability to communicate effectively in order to preserve their health.

Activity B

1. Answers will vary depending on the environment in which each student lives. Examples of constant or frequent loud noises that are involuntary include city traffic, subway noises, construction equipment, and thunder. Examples of ways to avoid these loud noises include wearing earplugs and changing one's transportation route.

2. Answers will vary but should include several voluntary sources of noise. Examples include radios, televisions, computer games, and loud musical instruments such as drums or electric guitars.

3. Answers will vary. Examples of voluntary activities that frequently involve loud noise include dances, motorcycle riding, and loud concerts. Examples of ways to protect one's hearing at these events include wearing earplugs, staying relatively far away from the source of the noise, and requesting that the noise be turned down.

Activity C

1. Answers may vary but should indicate an understanding that riding in separate cars creates more air pollution. Marcy could request that they take one car to the theater, or she could use a less damaging form of transportation, such as the bus or a bicycle.
2. Answers will vary but should demonstrate an ability to communicate effectively.
3. Answers will vary but should demonstrate an understanding that smoking is a source of air pollution. Students may suggest that Miguel stand his ground, have his own nonsmoking party, or designate a nonsmoking area.
4. Answers will vary but should demonstrate an ability to communicate effectively. Students may be aware that, if done correctly, compost heaps do not smell, can be well contained, and do not pose a health threat.
5. Answers will vary but should demonstrate an ability to communicate effectively and an understanding of the potentially damaging effects of loud music.

Activity D

While there are no correct answers to the questions in this activity, students' answers should reflect thorough research of community resources.

Chapter 26

Activity A

While there are no correct answers to the questions in this activity, students' answers should reflect a thorough interview.

Activity B

While there are no correct answers to the questions in this activity, students' answers should reflect thorough research of community resources.

Activity C

This activity is designed to help students practice communicating effectively in order to resist sales pressure. There are no correct answers to the questions in this activity.

Chapter 27

Activity A

1. Brands will vary depending on your geographic location. Fire extinguishers are classified according to three ratings: class A, class B, and class C. These ratings are determined by the type of fire the substance in the extinguishers will put out.
2. Answers will vary.
3. Class A fire extinguishers are used to put out ordinary combustibles, such as wood, paper, linen, rubber, and plastics. Class B fire extinguishers are used to put out flammable liquids, such as gasoline, oil, tar, grease, oil-based paints, lacquer, and flammable gas. Class C extinguishers are to be used on electrical wiring, fuse boxes, circuit breakers, machinery, and appliances. Using the wrong class of extinguisher on a fire may actually make the fire worse. Multipurpose, or ABC, extinguishers can be used in all three types of fires.
4. Because kitchens usually contain wood, grease, and electrical equipment, a multipurpose extinguisher would probably work best here. The type of fire extinguisher used in a basement would depend on the materials or equipment kept in the basement. In most cases, a class B extinguisher would be kept in the car to extinguish flammable liquids, such as gasoline or oil.

Chapter 28

Activity A

1. Answers may vary. Accept all reasonable answers. Students may correctly speculate that both an adult and a child are more likely to survive if the adult puts his or her own mask on first. If an adult puts the child's mask on first, the adult may faint before putting their own mask on. The child may be too young or too frightened to be able to help that adult. However, if the adult puts his or her own mask on first, then he or she will be alert enough to take care of a child.
2. Accept all reasonable answers. Answers will vary but should include examples for situations that would warrant protecting oneself first. For example, if four people were in a serious car

wreck and only one person was able to walk, that person should run to get help rather than attempt to stay and treat the other three. The injured people probably need more help than one untrained person can give.

Activity B

1. Supplies needed for treating wounds include sterile gauze or sanitary napkins, bandages, soap, clean water or a clean container for carrying water, a clean cloth, and a plastic bag (for possible avulsions).
2. Supplies needed for treating a fracture include clean bandages, clean cloth, and firm material for splints, such as pieces of wood or cardboard.
3. Supplies needed to avoid contact with a person's blood include rubber gloves and a clean cloth. Students may also be aware that mouth pieces are available to prevent contact during CPR.
4. Supplies needed to treat a head injury include a flashlight or other source of light to check for pupil dilation and soft objects to brace the head.
5. Supplies needed for treating burns include sterile dressings or clean cloth and clean water or a clean container for carrying water.
6. Supplies needed to apply pressure include sterile dressings, such as sterile gauze or sanitary napkins, and bandages to wrap around the dressing.

Activity C

While there are no correct answers to the questions in this activity, students' answers should reflect thorough research of community resources.

Activity D

1. Because they could have spinal injuries, the injured people should not be moved. Instead, they should be checked for airway obstruction, breathing, circulation, head injuries, and shock.
2. Deep puncture wounds are easily infected because they are harder to clean. The wound should be encouraged to bleed and then washed and covered to prevent infection. Students may also recommend following treatment by checking to make sure that the injured person is up-to-date on his or her tetanus shots.

3. As long as the abrasion has been rinsed, wiped to remove larger pieces of debris, and covered with a bandage, there is usually no need to remove the tiniest pieces of dirt.
4. A burn that is whitish in color is likely to be a third-degree burn. The fact that the victim says it does not hurt indicates that the burn was severe enough to destroy nerve endings. Applying ice could send the victim into shock. Instead, the burn should be rinsed gently with cool water, covered in sterile dressings, and elevated. The patient should then be monitored for symptoms of shock and encouraged to drink water until an emergency vehicle arrives.
5. If the cyclist has a head injury, elevating the legs could cause serious problems. The cyclist should be checked for a head injury before any other treatment is administered. The victim's ABCs should be checked and an emergency vehicle should be called.
6. If the snake was poisonous, running down a hill would accelerate the damaging effects of the venom. Instead, the wound should be washed with soap and water and the victim should lie still with the bite positioned lower than the level of the heart. A constricting band should be tied 2 to 4 inches above any swelling from the bite so that the band is snug enough to dimple the skin but just loose enough to allow for a pulse below the band. The uninjured friend should then go for help.
7. The injured ankle should be splinted and treated as if it were a fracture until the injury can be inspected by a doctor. No harm will come from treating a sprain as a fracture, but harm can be done by treating a fracture as a sprain.
8. Immersing the toes in hot water will cause more damage. Instead, the skin should be thawed in warm water only until the skin is flushed. The skin should not be rubbed and should be kept away from hot fires or stoves. Any blisters that form should be left unbroken.
9. These are the symptoms of heatstroke, a condition that cannot be resolved by "cooling off." The victim should be taken indoors or to a cool area, bathed in cool (not cold) water or damp cloths, and encouraged to drink small quantities of water. The victim should be transported to a hospital emergency room as quickly as possible.